1001 MATH PROBLEMS

3RD EDITION

1001 MATH PROBLEMS
3RD EDITION

LEARNINGEXPRESS ®

NEW YORK

Library of Congress Control Number: 2009926365

A copy of this book is on file with the Library of Congress.

Printed in the United States of America

9 8 7 6 5 4 3 2 1

Third Edition

ISBN 13: 978-1-57685-686-4

For more information or to place an order, contact LearningExpress at:
   2 Rector Street
   26th Floor
   New York, NY 10006

Or visit us at:
   www.learnatest.com

Contents ▶

▶ Introduction

This book—which can be used alone, in combination with the Learning-Express publication *Practical Math Success in 20 Minutes a Day*, or along with another basic math text of your choice—will give you practice in dealing with whole numbers, fractions, decimals, percentages, basic algebra, and basic geometry. It is designed for individuals working on their own, and for teachers or tutors helping students learn the basics. Practice on 1,001 math problems should help alleviate math anxiety, too!

Are you frightened of mathematics? You're not alone. By the time I was nine, I had developed a full-blown phobia. In fact, my most horrible moments in grade school took place right before an arithmetic test. My terror—and avoidance—lasted well into adulthood, until the day I landed a job with a social service agency and was given the task of figuring budgets, which involved knowing how to do percentages. I might just as well have been asked to decipher the strange squiggles incised on the nose-cone of an alien spaceship. I decided I'd better do something quick, so I went to a friend of mine, a fifth-grade teacher, and asked her to design a short course for me.

We met on Sundays for almost a year. She began each tutorial with a short lecture on the type of problem we would be working with, and then provided me with a yellow legal pad and a photocopied set of problems—and sat doing crossword puzzles while I labored. We worked our way up to geometry that way, and on into algebra.

"Mathematics works," she told me early on. "Don't ask why. Just do the problems. One day the light will dawn."

And it did finally! I'm proud to say I no longer have to pay someone to do my 1040 form for the IRS, and I don't squirm and make excuses when, at lunch with friends, I'm called on to figure the tip in my head. I even balance my checkbook now!

Learn by doing. It's an old lesson, tried and true. And it's the tool this book is designed to give you.

Of course, this method works for people who don't have math anxiety, too. Maybe you have simply forgotten a lot of what you learned about math because you haven't had to use it much. Or maybe you're a student tackling arithmetic, algebra, and geometry for the first time, and you just need more practice than your textbook gives you. Perhaps you're getting ready for an exam, and you just want to make sure your math skills are up to the task. Whatever your situation, you can benefit from the method of this book. That old maxim really is true: *Practice makes perfect.*

▶ An Overview of This Book

1001 Math Problems is divided into sections, each focusing on one kind of math:

Each section is subdivided into short sets of about 16 problems each, so as to make the whole project seem less overwhelming. You will begin with one or two sets of fairly simple non-word problems; later sets focus on word problems dealing with real-world situations.

In each section, you will find a few pre-algebra problems mixed in—problems that ask you to deal with *variables* (letters that stand for unknown numbers, such as x or y), *exponents* (those little numbers hanging above the other numbers, like 2^4), and the like. These problems are a warm-up for Section 5, Algebra. If they are too hard for you at first, just skip them. If you can answer them, you will be ahead of the game when you get to Section 5.

The most important learning tool in this book is not the problems, but the answers. At the back of the book, each answer is fully explained. After you finish a set, go to the back of the book and see how many questions you got right. But don't stop there: Look at the explanations for *all* the questions, both the ones you got right and the ones you got wrong. You will be learning by doing, and learning from your mistakes—the best way to learn *any* subject.

▶ How to Use This Book

Whether you are working alone or helping someone else brush up on their math, this book can help you improve math skills.

Working On Your Own

If you are working alone to brush up on the basics, you may want to use this book in combination with a basic text or with *Practical Math Success in 20 Minutes a Day.* It will be helpful to read a summary of the different kinds of fractions and how to convert fractions into another form, before tackling fraction problems. If you are fairly sure of your basic math skills, however, you can use this book by itself.

No matter how you decide to use the book, you will find it most helpful if you do **not** use a calculator for the majority of the problems, so as to prevent (or cure) "calculitis"—too much reliance on a calculator. (A special note is placed with the problems that will require you to use a calculator.)

Tutoring Others

This book will work very well in combination with almost any basic math text. You will probably find it most helpful to give the student a brief lesson in the particular operation they will be learning—whole numbers, fractions, decimals, percentages, algebra, or geometry—and then have him or her spend the remainder of the class or session actually doing problems. You will want to impress upon him or her the importance of learning by doing, and caution not to use a calculator so as to gain a better understanding of the operation in question.

▶ Additional Resources

You can find many other superb LearningExpress math review books at your local bookstore or library. Here are a few suggested titles:

- *Practical Math Success in 20 Minutes a Day.* Master real-world math skills in just four weeks! This invaluable skill-building guide makes challenging math subjects more accessible by tackling one small part of the larger topic and building upon that knowledge with each passing day. Complete each lesson in just 20 minutes a day to build skills in areas such as fractions, word problems, polygons and triangles, probability, and algebraic equations.
- Express Review Guides series. LearningExpress's Express Review Guides series includes a number of useful math titles. *Basic Math and Pre-Algebra,*

Algebra I, Algebra II, Fractions, Percentages & Decimals, and *Math Word Problems* are all designed with practical drills, numerous diagrams and charts, and tons of step-by-step practice exercises.

- *Math to the Max: 1200 Questions to Maximize Your Math Power.* With 1,200 total multiple-choice practice questions, this jam-packed review book covers areas such as basic math, algebra, geometry, and more. *Math to the Max* helps you master essential math skills to score higher on any standardized test, and includes detailed answer explanations to help you identify your strengths and weaknesses.
- *Office Financials Made Easy* supplies math problems to help you take stock of your basic office math skills. Whether you need a review of the fundamentals or instruction on more complex business math concepts, this guide will show you a wide range of applications, such as cash flow statements, depreciation, simple statistics, profit and loss statements, and payroll calculations.
- *Seeing Numbers.* This guide provides a new way to look at math with pictures, graphs, and diagrams. Designed especially for learners who have difficulty with conventional math techniques, *Seeing Numbers* gives you step-by-step instructions using pictures to help solve problems, sample questions with graphics illustrating important math concepts, and thorough answer explanations that walk you through each question.

1 ▶ Miscellaneous Math

The following section consists of 10 sets of miscellaneous math, including basic arithmetic questions and word problems with whole numbers. You will also see problems involving pre-algebra concepts such as negative numbers, exponents, and square roots (getting you ready for the algebra in Section 5). This section will provide a warm-up session before you move on to more difficult kinds of problems.

▶ **Set 1** (Answers begin on page 157.)

1. 8 + 6 =
   a. 2
   b. 16
   c. 14
   d. 48

2. 11 − 5 =
   a. 5
   b. 6
   c. 16
   d. 55

3. 9 ÷ 3 =
   a. 3
   b. 6
   c. 12
   d. 27

4. 4 × 4 =
   a. 8
   b. 0
   c. 1
   d. 16

5. 62 − 18 =
   a. 80
   b. 44
   c. 46
   d. 90

6. 93 + 7 =
   a. 100
   b. 99
   c. 86
   d. 101

7. 35 + 14 =
   a. 49
   b. 48
   c. 21
   d. 22

8. 56 + 44 =
   a. 12
   b. 99
   c. 100
   d. 18

9. 69 − 26 =
   a. 43
   b. 45
   c. 115
   d. 116

10. 82,404 − 202 =
   a. 82,606
   b. 82,202
   c. 82,102
   d. 82,001

11. 72 + 98 − 17 =
   a. 143
   b. 153
   c. 163
   d. 170

12. 376 − 360 + 337 =
   a. 663
   b. 553
   c. 453
   d. 353

13. $444 + 332 - 216 =$
   a. 312
   b. 450
   c. 560
   d. 612

14. $2,710 + 4,370 - 400 =$
   a. 6,780
   b. 6,680
   c. 6,580
   d. 6,480

15. $7,777 - 3,443 + 1,173 =$
   a. 5,507
   b. 5,407
   c. 5,307
   d. 5,207

16. $|7| =$
   a. −7
   b. 0.7
   c. 7
   d. 17

▶ **Set 2** (Answers begin on page 158.)

17. $(15 + 32)(56 − 39) =$
  a. 142
  b. 799
  c. 4,465
  d. 30

18. $72 ÷ 8 =$
  a. 9
  b. 8
  c. 7
  d. 6

19. $600 × 42 =$
  a. 2,520
  b. 20,500
  c. 25,020
  d. 25,200

20. $215 × 70 =$
  a. 1,505
  b. 10,550
  c. 15,050
  d. 15,500

21. $1,215 × 6 =$
  a. 7,290
  b. 7,920
  c. 7,092
  d. 7,029

22. What is the value of $56,515 ÷ 4$, rounded to the nearest whole number?
  a. 10,000
  b. 14,000
  c. 14,128
  d. 14,129

23. $|−9| =$
  a. −9
  b. −8
  c. 0
  d. 9

24. $(6 + 3) × 5 =$
  a. 21
  b. 45
  c. 90
  d. 33

25. $2,850 ÷ 190 =$
  a. 15
  b. 16
  c. 105
  d. 150

26. $62,035 ÷ 5 =$
  a. 1,247
  b. 12,470
  c. 12,407
  d. 13,610

27. Estimate the value of $6,302 ÷ 63$.
  a. 1
  b. 10
  c. 100
  d. 1,000

28. $4,563 × 45 =$
  a. 205,335
  b. 206,665
  c. 207,305
  d. 206,335

29. $12(84 - 5) - (3 \times 54) =$
   a. 786
   b. 796
   c. 841
   d. 54,000

30. $908 \times 276 =$
   a. 25,608
   b. 250,608
   c. 250,680
   d. 25,680

31. $(12 \times 6) + 18 =$
   a. 1,296
   b. 78
   c. 80
   d. 90

32. $(667 \times 2) + 133 =$
   a. 1,467
   b. 1,307
   c. 1,267
   d. 1,117

▶ **Set 3** (Answers begin on page 159.)

33. $(403 \times 163) - 678 =$
a. 11,003
b. 34,772
c. 65,011
d. 67,020

34. $(800 \div 40) + 142 =$
a. 126
b. 106
c. 162
d. 102

35. $604 - 204 \div 2 =$
a. 502
b. 301
c. 201
d. 101

36. $604 - (202 \div 2) =$
a. 201
b. 302
c. 402
d. 503

37. $52 \times 70 \times 4 =$
a. 12,560
b. 14,560
c. 11,560
d. 10,560

38. $17(8 \times 3) =$
a. 187
b. 139
c. 316
d. 408

39. Which of the following expressions are equal to 40,503?
a. $400 + 50 + 3$
b. $4,000 + 500 + 3$
c. $40,000 + 50 + 3$
d. $40,000 + 500 + 3$

40. Which of the following choices is divisible by both 7 and 8?
a. 42
b. 78
c. 112
d. 128

41. Which of the following numbers is represented by the prime factors $2 \times 3 \times 7$?
a. 21
b. 42
c. 84
d. 237

42. 184 is evenly divisible by
a. 46
b. 43
c. 41
d. 40

43. 4 pounds 14 ounces − 3 pounds 12 ounces =
a. 2 pounds 2 ounces
b. 1 pound 12 ounces
c. 1 pound 2 ounces
d. 14 ounces

44. 3 hours 20 minutes − 1 hour 48 minutes =
a. 5 hours 8 minutes
b. 4 hours 8 minutes
c. 2 hours 28 minutes
d. 1 hour 32 minutes

45. What is the mean of the following numbers: 76, 34, 78, and 56?

   a. 244

   b. 61

   c. 49

   d. 56

46. $39,824 \div 8 =$

   a. 4,798

   b. 4,897

   c. 4,978

   d. 4,987

47. 3 feet 6 inches + 5 feet 8 inches =

   a. 9 feet 2 inches

   b. 9 feet

   c. 8 feet 12 inches

   d. 8 feet

48. 1 hour 20 minutes + 3 hours 30 minutes =

   a. 4 hours

   b. 4 hours 20 minutes

   c. 4 hours 50 minutes

   d. 5 hours

▶ **Set 4** (Answers begin on page 159.)

49. What is the estimated product when 157 and 817 are rounded to the nearest hundred and multiplied?
   a. 160,000
   b. 180,000
   c. 16,000
   d. 80,000

50. $(-19) + 6 =$
   a. -25
   b. -13
   c. 0
   d. 13

51. $-4 \times (-9) =$
   a. 13
   b. -13
   c. 36
   d. -36

52. $-6 + (-6) =$
   a. -12
   b. 12
   c. 36
   d. -36

53. $7 + (-15) =$
   a. 22
   b. -22
   c. -8
   d. 8

54. $75 + (-75) =$
   a. 0
   b. 150
   c. -75
   d. 1

55. Which of the following choices is equivalent to $6 \times 6 \times 6$?
   a. 3×6
   b. 12×6
   c. 6^3
   d. 3^6

56. Which of the following choices is equivalent to 2^5?
   a. 7
   b. 10
   c. 16
   d. 32

57. Which of the following represents a composite number?
   a. 11
   b. 29
   c. 41
   d. 91

58. $19^2 =$
   a. 38
   b. 76
   c. 152
   d. 361

59. $10^5 \div 10^2 =$
   a. 1^3
   b. 10^3
   c. 10^7
   d. 10^{10}

60. $-4^2 =$
   a. -8
   b. 8
   c. -16
   d. 16

61. $-5^3 =$

   a. -15

   b. 15

   c. 125

   d. -125

62. $-11^2 =$

   a. 121

   b. -121

   c. -22

   d. 22

63. What is the square root of 64?

   a. 8

   b. 32

   c. 128

   d. $4,096$

64. Which of these equations is incorrect?

   a. $\sqrt{16} + \sqrt{3} = \sqrt{16 + 3}$

   b. $\sqrt{6} \times \sqrt{12} = \sqrt{6 \times 12}$

   c. Neither is incorrect.

   d. Both are incorrect.

► **Set 5** (Answers begin on page 160.)

65. What is the square root of 64?
   a. 16
   b. 12
   c. 8
   d. 6

66. The value of 5! is equal to
   a. 1
   b. 5
   c. 25
   d. 120

67. Alex bought 400 hot dogs for the school picnic. If they were contained in packages of eight hot dogs, how many total packages did he buy?
   a. 5
   b. 50
   c. 500
   d. 3,200

68. While preparing a dessert, Sue started by using 12 ounces of chocolate in her recipe. Later, she added 10 more ounces for flavor. What was the total amount of chocolate that Sue ended up using?
   a. 1 pound
   b. 1 pound 2 ounces
   c. 1 pound 4 ounces
   d. 1 pound 6 ounces

69. Dan rented two movies to watch last night. The first was 1 hour 40 minutes long, the second 1 hour 50 minutes long. How much time did it take for Dan to watch the two videos?
   a. 4.5 hours
   b. 3.5 hours
   c. 2.5 hours
   d. 1.5 hours

70. During a fund-raiser, each of the 35 members of a group sold candy bars. If each member sold an average of six candy bars, how many total bars did the group sell?
   a. 6
   b. 41
   c. 180
   d. 210

71. Roberta takes $58 with her on a shopping trip to the mall. She spends $18 on new shoes and another $6 on lunch. How much money does she have left after these purchases?
   a. $34
   b. $40
   c. $52
   d. $24

72. On a four-day trip Carrie drove 135 miles the first day, 213 miles the second day, 159 miles the third day, and 189 miles the fourth day. Which of the following choices is the best approximation of the total miles Carrie drove during the trip?
   a. 600
   b. 700
   c. 400
   d. 800

73. While bowling in a tournament, Jake and his friends had the following scores:

- Jake   189
- Charles and Max each scored   120
- Terry   95

What was the total score for Jake and his friends at the tournament?

   a. 404
   b. 504
   c. 524
   d. 526

74. The drivers at G & G trucking must report the mileage on their trucks each week. The mileage reading of Ed's vehicle was 20,907 at the beginning of one week, and 21,053 at the end of the same week. What was the total number of miles driven by Ed that week?

   a. 46 miles
   b. 145 miles
   c. 146 miles
   d. 1,046 miles

75. In a downtown department store, Angelo finds a woman's handbag and turns it in to the clerk in the Lost and Found Department. The clerk estimates that the handbag is worth approximately $150. Inside, she finds the following items:

- 1 leather makeup case valued at $65
- 1 vial of perfume, unopened,
  valued at $75
- 1 pair of earrings valued at $150
- Cash totaling $178

The clerk is writing a report to be submitted along with the found property. What should she write as the total value of the found cash and property?

   a. $468
   b. $608
   c. $618
   d. $718

76. A trash container, when empty, weighs 27 pounds. If this container is filled with a load of trash that weighs 108 pounds, what is the total weight of the container and its contents?

   a. 81 pounds
   b. 135 pounds
   c. 145 pounds
   d. 185 pounds

77. Mr. James Rossen is just beginning a computer consulting firm and has purchased the following equipment:

- three telephone sets, each costing $125
- two computers, each costing $1,300
- two computer monitors, each costing $950
- one printer costing $600
- one answering machine costing $50

Mr. Rossen is reviewing his finances. What should he write as the total value of the equipment he has purchased so far?

a. $3,025
b. $5,400
c. $5,525
d. $6,525

78. Mr. Richard Tupper is purchasing gifts for his family. So far he has purchased the following:

- three sweaters, each valued at $68
- one computer game valued at $75
- two bracelets, each valued at $43

Later, he returned one of the bracelets for a full refund and received a $10 rebate on the computer game. What is the total cost of the gifts after the refund and rebate?

a. $244
b. $312
c. $355
d. $365

79. Department regulations require trash collection trucks to have transmission maintenance every 13,000 miles. Truck #B-17 last had maintenance on its transmission at 12,398 miles. The mileage gauge now reads 22,003. How many more miles can the truck be driven before it must be brought in for transmission maintenance?

a. 3,395 miles
b. 4,395 miles
c. 9,003 miles
d. 9,605 miles

80. The city's bus system carries 1,200,000 people each day. How many people does the bus system carry each year? (one year = 365 days)

a. 3,288 people
b. 32,880 people
c. 43,800,000 people
d. 438,000,000 people

► **Set 6** (Answers begin on page 161.)

81. How many inches are there in four feet?
    a. 12 inches
    b. 36 inches
    c. 48 inches
    d. 52 inches

82. Dave is 46 years old, twice as old as Rajeeve. How old is Rajeeve?
    a. 30 years old
    b. 28 years old
    c. 23 years old
    d. 18 years old

83. Salesperson Rita drives 2,052 miles in six days, stopping at two towns each day. How many miles does she average between stops?
    a. 171 miles
    b. 342 miles
    c. 684 miles
    d. 1,026 miles

84. During the last week of track training, Shoshanna achieves the following times in seconds: 66, 57, 54, 54, 64, 59, and 59. Her three best times this week are averaged for her final score on the course. What is her final score?
    a. 57 seconds
    b. 55 seconds
    c. 59 seconds
    d. 61 seconds

85. A train must travel to a certain town in six days. The town is 3,450 miles away. How many miles must the train average each day to reach its destination?
    a. 500 miles
    b. 525 miles
    c. 550 miles
    d. 575 miles

86. On a certain day, the nurses at a hospital worked the following number of hours: Nurse Howard worked eight hours; Nurse Pease worked 10 hours; Nurse Campbell worked nine hours; Nurse Grace worked eight hours; Nurse McCarthy worked seven hours; and Nurse Murphy worked 12 hours. What is the average number of hours worked per nurse on this day?
    a. 7
    b. 8
    c. 9
    d. 10

87. A car uses 16 gallons of gas to travel 448 miles. How many miles per gallon does the car get?
    a. 22 miles per gallon
    b. 24 miles per gallon
    c. 26 miles per gallon
    d. 28 miles per gallon

88. A floppy disk shows 827,036 bytes free and 629,352 bytes used. If you delete a file of size 542,159 bytes and create a new file of size 489,986 bytes, how many free bytes will the floppy disk have?
    a. 577,179
    b. 681,525
    c. 774,863
    d. 879,209

89. A family's gas and electricity bill averages $80 a month for seven months of the year and $20 a month for the rest of the year. If the family's bills were averaged over the entire year, what would the monthly bill be?

a. $45
b. $50
c. $55
d. $60

90. If a vehicle is driven 22 miles on Monday, 25 miles on Tuesday, and 19 miles on Wednesday, what is the average number of miles driven each day?

a. 19 miles
b. 21 miles
c. 22 miles
d. 23 miles

91. A piece of gauze 3 feet 4 inches long was divided in five equal parts. How long was each part?

a. 1 foot 2 inches
b. 10 inches
c. 8 inches
d. 6 inches

92. Each sprinkler head on an athletic field sprays water at an average of 16 gallons per minute. If five sprinkler heads are flowing at the same time, how many gallons of water will be released in 10 minutes?

a. 80 gallons
b. 160 gallons
c. 800 gallons
d. 1,650 gallons

93. Use the following data to answer this question: Lefty keeps track of the length of each fish that he catches. Following are the lengths in inches of the fish that he caught one day:

12, 13, 8, 10, 8, 9, 17

What is the median fish length that Lefty caught that day?

a. 8 inches
b. 10 inches
c. 11 inches
d. 12 inches

94. If it takes two workers, working separately but at the same speed, 2 hours 40 minutes to complete a particular task, about how long will it take one worker, working at the same speed, to complete the same task alone?

a. 1 hour 20 minutes
b. 4 hours 40 minutes
c. 5 hours
d. 5 hours 20 minutes

95. A snack machine accepts only quarters. Candy bars cost 25¢, packages of peanuts cost 75¢, and cans of cola cost 50¢. How many quarters are needed to buy two candy bars, one package of peanuts, and one can of cola?

a. 8 quarters
b. 7 quarters
c. 6 quarters
d. 5 quarters

96. A street sign reads "Loading Zone 15 Minutes." If a truck pulls into this zone at 11:46 A.M., by what time must it leave?

a. 11:59 A.M.
b. 12:01 P.M.
c. 12:03 P.M.
d. 12:06 P.M.

► Set 7 (Answers begin on page 162.)

97. An elevator sign reads "Maximum weight 600 pounds." Which of the following may ride the elevator?
   a. three people: one weighing 198 pounds, one weighing 185 pounds, one weighing 200 pounds
   b. one person weighing 142 pounds with a load weighing 500 pounds
   c. one person weighing 165 pounds with a load weighing 503 pounds
   d. three people: one weighing 210 pounds, one weighing 101 pounds, one weighing 298 pounds

98. Darlene was hired to teach three identical math courses, which entailed being present in the classroom 48 hours altogether. At $35 per class hour, how much did Darlene earn for teaching one course?
   a. $105
   b. $560
   c. $840
   d. $1,680

99. Carmella and Mariah got summer jobs at the ice cream shop and were supposed to work 15 hours per week each for eight weeks. During that time, Mariah was ill for one week and Carmella took her shifts. How many hours did Carmella work during the eight weeks?
   a. 120 hours
   b. 135 hours
   c. 150 hours
   d. 185 hours

100. Jerry's Fish Market was shipped 400 pounds of cod packed into 20-pound crates. How many crates were needed for the shipment?
   a. 80 crates
   b. 40 crates
   c. 20 crates
   d. 10 crates

101. Which of the following is a translation of the statement "Twice the sum of six and four"?
   a. $2 + 6 + 4$
   b. $2 \times 6 + 4$
   c. $2(6 + 4)$
   d. $(2 \times 6) \times 4$

102. Rashaard went fishing six days in the month of June. He caught eleven, four, zero, five, four, and six fish respectively. On the days that Rashaard fished, what was his average catch?
   a. 4
   b. 5
   c. 6
   d. 7

Answer question 103 on the basis of the following paragraph.

Basic cable television service, which includes 16 channels, costs $15 a month. The initial labor fee to install the service is $25. A $65 deposit is required but will be refunded within two years if the customer's bills are paid in full. Other cable services may be added to the basic service: The movie channels service is $9.40 a month; the news channels are $7.50 a month; the arts channels are $5.00 a month; and the sports channels are $4.80 a month.

103. A customer's first bill after having cable television installed totaled $112.50. This customer chose basic cable and one additional cable service. Which additional service was chosen?
   a. the news channels
   b. the movie channels
   c. the arts channels
   d. the sports channels

104. An army food supply truck can carry three tons. A breakfast ration weighs 12 ounces, and the other two daily meals weigh 18 ounces each. Assuming each soldier gets 3 meals per day, on a 10-day trip, how many soldiers can be supplied by one truck?
   a. 100
   b. 150
   c. 200
   d. 320

105. Meda arrived at work at 8:14 A.M., and Kirstin arrived at 9:12 A.M. How long had Meda been at work when Kirstin got there?
   a. 1 hour 8 minutes
   b. 1 hour 2 minutes
   c. 58 minutes
   d. 30 minutes

106. A clerk can process 26 forms per hour. If 5,600 forms must be processed in an eight-hour day, how many clerks must you hire for that day?
   a. 24
   b. 25
   c. 26
   d. 27

107. How many different meals can be ordered from a restaurant if there are three choices of soup, five choices of entrées, and two choices of dessert if a meal consists of a soup, entrée, and dessert?
   a. 10
   b. 15
   c. 30
   d. 60

108. Which of the following best represents the following sentence? Rachel (R) had three apples and ate one.
   a. $R = 3 - 1$
   b. $3 - 2 = R$
   c. $R = 3 \times 2$
   d. $3R - 2$

109. A uniform requires 4 square yards of cloth. To produce uniforms for 84,720 troops, how much cloth is required?
   a. 21,180 square yards
   b. 21,880 square yards
   c. 338,880 square yards
   d. 340,880 square yards

110. Marty left his workplace at 5:16 P.M. on Thursday and returned at 7:58 A.M. on Friday. How much time elapsed between the time Marty left work on Thursday to the time he returned on Friday?
   a. 2 hours 42 minutes
   b. 13 hours 42 minutes
   c. 14 hours 42 minutes
   d. 14 hours 52 minutes

111. Evaluate the expression: $|-14| + -5$
   a. -19
   b. 19
   c. 9
   d. -9

112. The cost of a certain type of fruit is displayed in the following table.

Weight (in lbs.)	Cost (in dollars)
4 lbs.	$1.10
5 lbs.	$1.74

Based on the table, estimate the cost for 4 pounds 8 ounces of the same type of fruit.
   a. $1.24
   b. $1.32
   c. $1.35
   d. $1.42

▶ **Set 8** (Answers begin on page 163.)

113. Larry buys three puppies at the Furry Friends Kennel for a total cost of $70. Two of the puppies are on sale for $15 apiece. How much does the third puppy cost?

    a. $55

    b. $40

    c. $30

    d. $25

Use the following table to answer question 114.

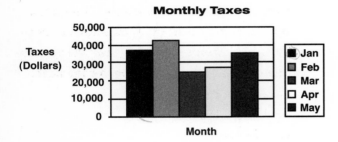

114. Approximately what were the total taxes collected for January, February, and April?

    a. $78,000

    b. $98,000

    c. $105,000

    d. $115,000

115. If Linda purchases an item that costs $30 or less, she will pay with cash.

If Linda purchases an item that costs between $30 and $70, she will pay with a check.

If Linda purchases an item that costs $70 or greater, she will use a credit card.

If Linda recently paid for a certain item using a check, which of the following statements could be true?

    a. The item cost $80.

    b. If the item had cost $20 more, she would have paid with cash.

    c. The item cost at least $70.

    d. The item cost more than $25.

116. What is the next number in the following series?

    3   16   6   12   12   8   __

    a. 4

    b. 15

    c. 20

    d. 24

117. Five people in Sonja's office are planning a party. Sonja will buy a loaf of French bread ($3 per loaf) and a platter of cold cuts ($23). Barbara will buy the soda ($1 per person) and two boxes of crackers ($2 per box). Mario and Rick will split the cost of two packages of Cheese Doodles ($1 per package). Danica will supply a package of five paper plates ($4 per package). How much more will Sonja spend than the rest of the office put together?

    a. $14

    b. $13

    c. $12

    d. $11

118. In the music department at a school, a music teacher counted the musical instruments and supplies in storage. There was:

- one violin valued at $1,200
- two violin bows, each valued at $350
- three music stands, each valued at $55
- one trumpet valued at $235

In addition, there were a number of supplies totaling $125 and some sheet music worth $75. What was the total value of the musical supplies and instruments in storage?
   a. $2,040
   b. $2,500
   c. $3,040
   d. $3,500

119. What is the value of the expression $5(4^0)$?
   a. 0
   b. 1
   c. 5
   d. 20

120. If a population of yeast cells grows from 10 to 320 in a period of five hours, what is the rate of growth?
   a. It doubles its numbers every hour.
   b. It triples its numbers every hour.
   c. It doubles its numbers every two hours.
   d. It triples its numbers every two hours.

121. The number of red blood corpuscles in one cubic millimeter is about 5,000,000, and the number of white blood corpuscles in one cubic millimeter is about 8,000. What, then, is the ratio of white blood corpuscles to red blood corpuscles?
   a. 1:625
   b. 1:40
   c. 4:10
   d. 5:1,250

122. George has made a vow to jog for an average of one hour daily five days a week. He cut his workout short on Wednesday by 40 minutes, but was able to make up 20 minutes on Thursday and 13 minutes on Friday. How many minutes of jogging did George lose for the week?
   a. 20 minutes
   b. 13 minutes
   c. 7 minutes
   d. 3 minutes

123. Which of the following is equivalent to $2\sqrt{6}$?
   a. $\sqrt{24}$
   b. $6\sqrt{2}$
   c. $12\sqrt{2}$
   d. $\sqrt{12}$

124. Which of the following is an irrational number?
   a. 0.333333…
   b. $\frac{1}{2}$
   c. 10.489
   d. 0.101101110…

125. Joni is 5 feet 11 inches tall, and Pierre is 6 feet 5 inches tall. How much taller is Pierre than Joni?
- **a.** 1 foot 7 inches
- **b.** 1 foot
- **c.** 7 inches
- **d.** 6 inches

126. Marcia is 10 years older than Fred, who is 16. How old is Marcia?
- **a.** 6 years old
- **b.** 20 years old
- **c.** 26 years old
- **d.** 30 years old

127. Which of the following numbers can be divided evenly by 19?
- **a.** 54
- **b.** 63
- **c.** 76
- **d.** 82

128. Find the mode of the following set of numbers:

8 4 6 12 9 9 4 4
- **a.** 4
- **b.** 6
- **c.** 7
- **d.** 9

▶ **Set 9** (Answers begin on page 165.)

129. Emilio is 1 year 7 months old, and Brooke is 2 years 8 months old. How much older is Brooke than Emilio?

a. 1 year 1 month

b. 2 years

c. 1 month

d. 1 year 2 months

130. Fifth graders Kara and Rani both have lemonade stands. Kara sells her lemonade at five cents a glass, and Rani sells hers at seven cents a glass. Kara sold 17 glasses of lemonade today, and Rani sold 14 glasses. Who made the most money and by what amount?

a. Kara by 13 cents

b. Rani by 13 cents

c. Kara by 85 cents

d. Rani by 98 cents

131. A train travels 300 miles in six hours. If it was traveling at a constant speed the entire time, what was the speed of the train?

a. 50 miles per hour

b. 60 miles per hour

c. 180 miles per hour

d. 1,800 miles per hour

132. If Rita can run around the block five times in 20 minutes, how many times can she run around the block in one hour?

a. 10

b. 15

c. 50

d. 100

133. Which of the following number sentences is true?

a. 4 feet > 3 feet

b. 7 feet < 6 feet

c. 5 feet > 6 feet

d. 3 feet < 2 feet

134. Which of the following is a prime number?

a. 6

b. 9

c. 11

d. 27

135. Which of the following is a translation of the following sentence? Salwa (S) is ten years older than Roland (R).

a. $10 + S = R$

b. $S + R = 10$

c. $R - 10 = S$

d. $S = R + 10$

136. How many ways can four students line up in a line, if the order matters?

a. 4

b. 8

c. 16

d. 24

137. At the movies, Lucinda bought food for herself and her friend Rae, including: one box of popcorn to share at $5 a box, one box of Junior Mints for each of them at $2 a box, and one soft drink for each at $3 apiece. Rae bought a ticket for each at $7 apiece. Who spent the most money and by how much?

a. Rae by $3

b. Rae by $7

c. Lucinda by $1

d. Lucinda by $2

138. One colony of bats consumes 36 tons of mosquitoes per year. At that rate, how many pounds of mosquitoes does the same colony consume in a month?
   a. 36,000 pounds
   b. 12,000 pounds
   c. 6,000 pounds
   d. 3,000 pounds

139. Which of the following answer choices is equivalent to 10^4?
   a. $10 \times 10 \times 10 \times 10$
   b. 10×4
   c. $(10 + 4) \times 10$
   d. $10 + 4$

140. How many acres are contained in a parcel 121 feet wide and 240 yards deep? (one acre = 43,560 square feet)
   a. 1 acre
   b. $1\frac{1}{2}$ acres
   c. 2 acres
   d. $2\frac{1}{2}$ acres

141. Land consisting of a quarter section is sold for $1,850 per acre (one quarter section = 160 acres). The total sale price is
   a. $296,000
   b. $592,000
   c. $1,184,000
   d. $1,850,000

142. The dwarf planet Pluto is estimated at a mean distance of 3,666 million miles from the sun. The planet Mars is estimated at a mean distance of 36 million miles from the Sun. How much closer to the Sun is Mars than Pluto?
   a. 36,300,000 million miles
   b. 36,300 million miles
   c. 3,630 million miles
   d. 363 million miles

143. A rectangular tract of land measures 860 feet by 560 feet. Approximately how many acres is this? (one acre = 43,560 square feet)
   a. 12.8 acres
   b. 11.06 acres
   c. 10.5 acres
   d. 8.06 acres

144. A dormitory now houses 30 men and allows 42 square feet of space per man. If five more men are put into this dormitory, how much less space will each man have?
   a. 5 square feet
   b. 6 square feet
   c. 7 square feet
   d. 8 square feet

▶ **Set 10** (Answers begin on page 166.)

145. On a particular morning the temperature went up 1° every two hours. If the temperature was 53° at 5 A.M., at what time was it 57°?
   a. 7 A.M.
   b. 8 A.M.
   c. 12 P.M.
   d. 1 P.M.

146. What is the value of $16^{\frac{1}{2}}$?
   a. 2
   b. 4
   c. 8
   d. 32

147. Find the median of the following group of numbers: 14 12 20 22 14 16
   a. 12
   b. 14
   c. 15
   d. 16

148. $16\sqrt{2} - 4\sqrt{2} =$
   a. 12
   b. $12\sqrt{2}$
   c. $12 - \sqrt{2}$
   d. 20

149. Which of the following is equivalent to $\sqrt{20}$?
   a. $2\sqrt{5}$
   b. $4\sqrt{5}$
   c. $10\sqrt{2}$
   d. 20

150. Which of the following numbers is divisible by six?
   a. 232
   b. 341
   c. 546
   d. 903

151. Which of the following best represents the following statement? Patricia (P) has four times the number of marbles Sean (S) has.
   a. P = S + 4
   b. S = P − 4
   c. P = 4S
   d. S = 4P

152. Write ten thousand, four hundred forty-seven in numerals.
   a. 10,499,047
   b. 104,447
   c. 10,447
   d. 1,047

153. Write ten million, forty-three thousand, seven hundred three in numerals.
   a. 143,703
   b. 1,043,703
   c. 10,043,703
   d. 10,430,703

154. John has started an egg farm. His chickens produce 480 eggs per day, and his eggs sell for $2 a dozen. How much does John make on eggs per week? (one week = seven days)
   a. $480
   b. $500
   c. $560
   d. $600

155. What is the value of 5^{-2}?

a. -10

b. -25

c. $\frac{1}{10}$

d. $\frac{1}{25}$

156. $|4 - 15| =$

a. -11

b. 11

c. -19

d. 19

Use the following table to answer question 157.

**PRODUCTION OF FARM-IT TRACTORS
FOR THE MONTH OF APRIL**

Factory	April Output
Dallas	450
Houston	425
Lubbock	
Amarillo	345
TOTAL	1,780

157. What was Lubbock's production in the month of April?

a. 345

b. 415

c. 540

d. 560

158. Fifty-four students are to be separated into six groups of equal size. How many students are in each group?

a. 8

b. 9

c. 10

d. 12

Use the following table to answer question 159.

STEVE'S BIRD-WATCHING PROJECT

Day	Number of Raptors Seen
Monday	
Tuesday	7
Wednesday	12
Thursday	11
Friday	4
MEAN	8

159. This table shows the data Steve collected while watching birds for one week. How many raptors did Steve see on Monday?

a. 6

b. 7

c. 8

d. 10

Use the following table to answer question 160.

**DISTANCE TRAVELED FROM
CHICAGO WITH RESPECT TO TIME**

Time (hours)	Distance from Chicago (miles)
1	60
2	120
3	180
4	240

160. A train moving at a constant speed leaves Chicago for Los Angeles [at time $t = 0$]. If Los Angeles is 2,000 miles from Chicago, which of the following equations describes the distance (D) from Los Angeles at any time t?

a. $D(t) = 60t - 2{,}000$

b. $D(t) = 60t$

c. $D(t) = 2{,}000 - 60t$

d. $D(t) = 2{,}000 + 60t$

2 ▶ Fractions

The following 10 sets of fraction problems will provide you with exercises in how to convert fractions and how to do arithmetic problems that involve fractions. (Sections 3 and 4 deal with decimals and percentages, which also show fractions of a number and are, therefore, called something else for clarity.) In order to understand arithmetic, it is important to practice and become comfortable with fractions and how they work.

You will start off with questions that just deal with numbers. After you've had a chance to practice your basic fraction skills, you can move on to some word problems involving fractions.

▶ **Set 11** (Answers begin on page 168.)

161. Name the fraction that indicates the shaded part of the following figure.

   a. $\frac{1}{4}$

   b. $\frac{1}{2}$

   c. $\frac{2}{3}$

   d. $\frac{3}{4}$

162. Name the fraction that indicates the shaded part of the following figure.

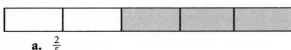

   a. $\frac{2}{5}$

   b. $\frac{3}{5}$

   c. $\frac{5}{3}$

   d. $\frac{5}{2}$

163. Which of the following represents $\frac{42}{56}$ in lowest terms?

   a. $\frac{21}{28}$

   b. $\frac{6}{8}$

   c. $\frac{3}{4}$

   d. $\frac{7}{8}$

164. Express the fraction $\frac{54}{108}$ in lowest terms.

   a. $\frac{27}{54}$

   b. $\frac{9}{18}$

   c. $\frac{3}{6}$

   d. $\frac{1}{2}$

165. $\frac{5}{9} - \frac{2}{9} =$

   a. $\frac{7}{9}$

   b. $\frac{3}{18}$

   c. $\frac{1}{3}$

   d. $\frac{2}{3}$

166. $\frac{1}{4} + \frac{3}{16} + \frac{7}{8} =$

   a. $1\frac{5}{16}$

   b. $\frac{11}{28}$

   c. $\frac{7}{16}$

   d. $1\frac{7}{16}$

167. $5\frac{1}{3} + 7 + 2\frac{1}{3} =$

   a. $14\frac{1}{3}$

   b. $\frac{1}{3}$

   c. 14

   d. $14\frac{2}{3}$

168. $8\frac{9}{13} - 5\frac{2}{13} =$

   a. $3\frac{7}{13}$

   b. $3\frac{11}{13}$

   c. $13\frac{7}{13}$

   d. $13\frac{11}{13}$

169. $\frac{1}{5} \times \frac{4}{7} =$

   a. $\frac{5}{12}$

   b. $\frac{4}{35}$

   c. $\frac{1}{35}$

   d. $\frac{2}{17}$

170. $\frac{5}{12} - \frac{7}{18} =$

   a. $\frac{1}{36}$

   b. $\frac{12}{36}$

   c. $\frac{12}{30}$

   d. $\frac{2}{6}$

171. $78\frac{2}{3} - 10\frac{4}{7} =$

   a. $68\frac{2}{21}$

   b. $69\frac{4}{21}$

   c. $68\frac{4}{21}$

   d. $58\frac{6}{10}$

172. $-\frac{5}{12} \div -\frac{1}{6} =$

a. $-\frac{5}{72}$

b. $-2\frac{1}{2}$

c. $2\frac{1}{2}$

d. $\frac{5}{72}$

173. $40 \div 2\frac{1}{2} =$

a. 4

b. 16

c. 80

d. 100

174. $\frac{1}{3} \div \frac{2}{7} =$

a. $2\frac{4}{5}$

b. $1\frac{1}{6}$

c. $2\frac{1}{7}$

d. $1\frac{1}{5}$

175. $\frac{7}{8} - \frac{3}{5} =$

a. $\frac{11}{40}$

b. $1\frac{1}{3}$

c. $\frac{1}{10}$

d. $1\frac{19}{40}$

176. $4\frac{3}{5} + -1\frac{2}{5} =$

a. $3\frac{1}{5}$

b. 4

c. $3\frac{1}{2}$

d. $5\frac{1}{5}$

▶ **Set 12** (Answers begin on page 169.)

177. $76\frac{1}{2} + 11\frac{5}{6} =$
   a. $87\frac{1}{2}$
   b. $88\frac{1}{3}$
   c. $88\frac{5}{6}$
   d. $89\frac{1}{6}$

178. $35\frac{7}{9} - 20\frac{2}{9} =$
   a. $15\frac{9}{9}$
   b. 16
   c. $15\frac{5}{9}$
   d. $15\frac{4}{9}$

179. $43\frac{2}{3} + 36\frac{3}{9} =$
   a. 100
   b. 90
   c. 80
   d. 70

180. $\frac{4}{7} - \frac{1}{3} =$
   a. $\frac{5}{10}$
   b. $\frac{3}{4}$
   c. $\frac{4}{21}$
   d. $\frac{5}{21}$

181. $\frac{5}{6} + \frac{3}{8} =$
   a. $\frac{8}{14}$
   b. $\frac{2}{14}$
   c. $\frac{11}{24}$
   d. $1\frac{5}{24}$

182. $\frac{7}{9} \times \frac{9}{7} =$
   a. 1
   b. $1\frac{1}{9}$
   c. $\frac{1}{63}$
   d. $\frac{1}{9}$

183. $-\frac{1}{6} \div -\frac{1}{12} =$
   a. -2
   b. 2
   c. $\frac{1}{72}$
   d. $-\frac{1}{72}$

184. $7\frac{3}{5} \times \frac{4}{9} =$
   a. $7\frac{4}{15}$
   b. $3\frac{1}{15}$
   c. $7\frac{1}{2}$
   d. $3\frac{17}{45}$

185. $\frac{2}{6} \div 2 =$
   a. $\frac{4}{6}$
   b. $\frac{3}{6}$
   c. $\frac{2}{6}$
   d. $\frac{1}{6}$

186. $2\frac{1}{4} \div 2\frac{4}{7} =$
   a. $\frac{9}{14}$
   b. $\frac{7}{8}$
   c. $1\frac{2}{7}$
   d. $5\frac{11}{14}$

187. $\frac{5}{12} \times \frac{1}{6} \times \frac{2}{3} =$
   a. $\frac{10}{12}$
   b. $\frac{5}{6}$
   c. $\frac{5}{108}$
   d. $\frac{5}{216}$

188. $1\frac{1}{2} \div 1\frac{5}{13} =$
   a. $1\frac{3}{10}$
   b. $1\frac{1}{12}$
   c. $2\frac{1}{13}$
   d. $3\frac{9}{10}$

189. $2\frac{1}{3} \times 1\frac{1}{14} \times 1\frac{4}{5} =$

    a. $1\frac{7}{18}$

    b. $2\frac{1}{2}$

    c. $3\frac{6}{7}$

    d. $4\frac{1}{2}$

190. $\frac{7}{9} \times \frac{4}{5} =$

    a. $\frac{28}{45}$

    b. $\frac{11}{14}$

    c. $\frac{35}{36}$

    d. $\frac{3}{4}$

191. $2\frac{1}{4} \div \frac{2}{3} =$

    a. $\frac{8}{27}$

    b. $1\frac{1}{2}$

    c. $3\frac{3}{8}$

    d. $3\frac{1}{2}$

192. $\frac{4}{7} \div \frac{8}{17} =$

    a. $3\frac{7}{119}$

    b. $\frac{32}{119}$

    c. $1\frac{3}{14}$

    d. $2\frac{3}{14}$

▶ **Set 13** (Answers begin on page 169.)

193. $8 \times \frac{1}{5} =$

a. $1\frac{1}{5}$

b. $1\frac{3}{5}$

c. $\frac{5}{8}$

d. $\frac{1}{40}$

194. $7 \times \frac{2}{5} =$

a. $\frac{1}{5}$

b. $17\frac{1}{2}$

c. $2\frac{4}{5}$

d. $2\frac{2}{5}$

195. $\frac{1}{6} \div 4\frac{5}{8} =$

a. $\frac{4}{5}$

b. $27\frac{3}{4}$

c. $\frac{37}{48}$

d. $\frac{4}{111}$

196. $3\frac{4}{7} \div \frac{1}{8} =$

a. $28\frac{4}{7}$

b. $28\frac{5}{7}$

c. 29

d. $29\frac{4}{7}$

197. $1\frac{1}{2} \div 2\frac{1}{4} =$

a. $\frac{2}{3}$

b. $1\frac{1}{2}$

c. $2\frac{3}{8}$

d. $2\frac{1}{8}$

198. $2\frac{1}{3} \times 5\frac{1}{2} \times \frac{3}{11} =$

a. $3\frac{1}{2}$

b. $7\frac{1}{2}$

c. $10\frac{1}{22}$

d. $10\frac{1}{6}$

199. $6\frac{2}{3} \div \frac{1}{3} =$

a. $6\frac{2}{9}$

b. 20

c. $20\frac{2}{9}$

d. $2\frac{2}{9}$

200. $\frac{2}{3} \times 5\frac{1}{9} =$

a. $3\frac{11}{27}$

b. $5\frac{2}{27}$

c. $5\frac{11}{27}$

d. $3\frac{2}{27}$

201. $-\frac{9}{7} - \frac{5}{7} =$

a. $-\frac{4}{7}$

b. $\frac{4}{7}$

c. -2

d. 2

202. $\frac{3}{4} \times \frac{16}{15} =$

a. $\frac{5}{4}$

b. $\frac{4}{5}$

c. $1\frac{4}{5}$

d. $1\frac{1}{4}$

203. $5 \div \frac{5}{8} =$

a. 8

b. $\frac{1}{8}$

c. $3\frac{1}{8}$

d. $\frac{5}{8}$

204. $6\frac{2}{9} - \frac{1}{6} =$

a. $6\frac{1}{18}$

b. $6\frac{5}{27}$

c. $6\frac{1}{3}$

d. $5\frac{8}{9}$

205. $9\frac{3}{7} + 4\frac{2}{5} =$
- **a.** $13\frac{5}{12}$
- **b.** $13\frac{3}{4}$
- **c.** $13\frac{29}{35}$
- **d.** $13\frac{37}{52}$

206. $3\frac{5}{6} \times 4\frac{2}{3} =$
- **a.** $17\frac{8}{9}$
- **b.** $12\frac{7}{18}$
- **c.** $16\frac{2}{3}$
- **d.** $13\frac{3}{5}$

207. Which of the following is between $\frac{1}{3}$ and $\frac{1}{4}$?
- **a.** $\frac{1}{5}$
- **b.** $\frac{2}{3}$
- **c.** $\frac{2}{5}$
- **d.** $\frac{2}{7}$

208. Change $\frac{55}{6}$ to a mixed number.
- **a.** $8\frac{1}{6}$
- **b.** $9\frac{1}{6}$
- **c.** $9\frac{1}{55}$
- **d.** $9\frac{6}{55}$

▶ **Set 14** (Answers begin on page 170.)

209. Which of the following is the equivalent of $\frac{18}{45}$?
   a. 0.45
   b. 0.5
   c. 0.42
   d. 0.4

210. Which of the following has the greatest value?
   a. $\frac{7}{8}$
   b. $\frac{3}{4}$
   c. $\frac{2}{3}$
   d. $\frac{5}{6}$

211. Which of the following has the smallest value?
   a. $\frac{3}{5}$
   b. $\frac{8}{15}$
   c. $\frac{17}{30}$
   d. $\frac{2}{3}$

212. If each of the following represents the diameter of a circle, which is the smallest diameter?
   a. $\frac{17}{20}$ inch
   b. $\frac{3}{4}$ inch
   c. $\frac{5}{6}$ inch
   d. $\frac{7}{10}$ inch

213. What is the reciprocal of $3\frac{7}{8}$?
   a. $\frac{31}{8}$
   b. $\frac{8}{31}$
   c. $\frac{8}{21}$
   d. $-\frac{31}{8}$

214. What is the reciprocal of $4\frac{3}{5}$?
   a. $\frac{23}{4}$
   b. $\frac{4}{23}$
   c. $\frac{5}{23}$
   d. $\frac{23}{5}$

215. Change $\frac{160}{40}$ to a whole number.
   a. 16
   b. 10
   c. 8
   d. 4

216. Change this improper fraction to a mixed number: $\frac{31}{3}$.
   a. 10
   b. $10\frac{1}{3}$
   c. $10\frac{1}{2}$
   d. $11\frac{1}{3}$

217. Change this mixed number to an improper fraction: $5\frac{1}{2}$.
   a. $\frac{11}{2}$
   b. $\frac{5}{1}$
   c. $\frac{7}{2}$
   d. $\frac{5}{2}$

218. $\frac{3}{5} - \frac{1}{4} =$
   a. 2
   b. $\frac{1}{10}$
   c. $\frac{3}{4}$
   d. $\frac{7}{20}$

219. Which of the following is an improper fraction?
   a. $\frac{22}{60}$
   b. $\frac{66}{22}$
   c. $\frac{90}{100}$
   d. $\frac{1,000}{2,600}$

220. Mario has finished 35 out of 45 of his test questions. Which of the following fractions of the test does he have left?

a. $\frac{2}{9}$

b. $\frac{7}{9}$

c. $\frac{4}{5}$

d. $\frac{3}{5}$

221. Joe gave $\frac{1}{2}$ of his sandwich to Ed at lunchtime and ate $\frac{1}{3}$ of it himself. How much of the sandwich did he have left?

a. $\frac{1}{6}$

b. $\frac{3}{5}$

c. $\frac{4}{5}$

d. $\frac{5}{6}$

222. Kevin is buying fabric for new curtains. There are three windows, each 35 inches wide. Kevin needs to buy fabric equal to $2\frac{1}{2}$ times the total width of the windows. How much fabric should he buy?

a. $262\frac{1}{2}$ inches

b. $175\frac{1}{3}$ inches

c. $210\frac{3}{4}$ inches

d. $326\frac{1}{4}$ inches

223. Katie and her family ordered a pizza for dinner and ate $\frac{3}{4}$ of it. The next day, Katie ate $\frac{1}{2}$ of what was leftover for lunch. What fraction of the original pizza did Katie eat for lunch?

a. $\frac{1}{8}$

b. $\frac{1}{6}$

c. $\frac{1}{4}$

d. $\frac{1}{2}$

224. Molly needs $\frac{5}{8}$ cup of diced onion for a recipe. After chopping all the onion she has, Molly has $\frac{3}{5}$ cup of chopped onion. How much more chopped onion does she need?

a. $\frac{1}{8}$ cup

b. $\frac{1}{5}$ cup

c. $\frac{1}{40}$ cup

d. $\frac{1}{60}$ cup

▶ **Set 15** (Answers begin on page 171.)

225. Hans has $5\frac{1}{2}$ pounds of sugar. He wants to make cookies for his son's kindergarten class. The cookie recipe calls for $\frac{2}{3}$ pound of sugar per dozen cookies. How many dozen cookies can he make?

a. $6\frac{1}{3}$ dozen

b. $7\frac{1}{5}$ dozen

c. $8\frac{1}{4}$ dozen

d. $9\frac{1}{2}$ dozen

226. John bought two pounds of butter to make cookies. If he used $\frac{1}{2}$ pound for chocolate chip cookies, $\frac{1}{8}$ pound for peanut butter cookies, and $\frac{2}{3}$ pound for sugar cookies, what part of the original two pounds is left?

a. $\frac{4}{13}$

b. $\frac{22}{13}$

c. $\frac{17}{24}$

d. $\frac{31}{24}$

227. Vonda is making a mosaic. Each tiny piece of glass in the artwork is $1\frac{1}{4}$ inch by $1\frac{3}{8}$ inch. What is the area of each piece?

a. $1\frac{23}{32}$ square inches

b. $1\frac{21}{22}$ square inches

c. $1\frac{23}{25}$ square inches

d. $1\frac{29}{31}$ square inches

228. Dan purchases $6\frac{1}{2}$ pounds of potato chips for a party. If there are a total of eight people at the party, how many pounds of chips does each person get?

a. $\frac{13}{16}$ pound

b. $1\frac{1}{4}$ pounds

c. 2 pounds

d. $2\frac{1}{4}$ pounds

229. Marilyn has $17\frac{3}{4}$ feet of wallpaper border. Each of the four walls in her bathroom is nine feet long. How much more wallpaper border does Marilyn need?

a. $17\frac{3}{4}$ feet

b. $16\frac{1}{2}$ feet

c. $18\frac{1}{4}$ feet

d. $19\frac{1}{2}$ feet

230. A recipe calls for all the liquid ingredients to be mixed together: $2\frac{1}{4}$ cups water, $4\frac{5}{8}$ cups chicken stock, and $\frac{1}{2}$ cup honey. How many cups of liquid are in the recipe?

a. $6\frac{7}{8}$ cups

b. $7\frac{1}{4}$ cups

c. $7\frac{3}{8}$ cups

d. $7\frac{3}{4}$ cups

231. A loaf of bread has 35 slices. Ann eats eight slices, Betty eats six slices, Carl eats five slices, and Derrick eats nine slices. What fraction of the loaf is left?

a. $\frac{2}{11}$

b. $\frac{1}{9}$

c. $\frac{2}{7}$

d. $\frac{1}{5}$

232. Frances wants to run $2\frac{1}{3}$ miles every day. Today she has gone $\frac{7}{8}$ mile. How much farther does she have to go?

   a. $1\frac{11}{24}$ miles

   b. $1\frac{1}{3}$ miles

   c. $1\frac{41}{50}$ miles

   d. $1\frac{307}{308}$ miles

233. Ribbon in a craft store costs $0.75 per yard. Vernon needs to buy $7\frac{1}{3}$ yards. How much will it cost?

   a. $7.33

   b. $6.95

   c. $5.50

   d. $4.25

234. Linda needs to read 14 pages for her history class, 26 pages for English, 12 pages for civics, and 28 pages for biology. She has read $\frac{1}{6}$ of the entire number of pages. How many pages has she read?

   a. 80

   b. $13\frac{1}{3}$

   c. $48\frac{1}{2}$

   d. 17

235. Ted has to write a $5\frac{1}{2}$-page paper. He's finished $3\frac{1}{3}$ pages. How many pages does he have left to write?

   a. $1\frac{3}{5}$

   b. $1\frac{7}{8}$

   c. $2\frac{2}{3}$

   d. $2\frac{1}{6}$

236. Maria made $331.01 last week. She worked $39\frac{1}{2}$ hours. What is her hourly wage?

   a. $8.28

   b. $8.33

   c. $8.38

   d. $8.43

237. Virgil ate $\frac{3}{7}$ of a chocolate chip cookie; Aristotle ate $\frac{1}{3}$ of the same cookie. How much of the cookie is left?

   a. $\frac{1}{3}$

   b. $\frac{3}{7}$

   c. $\frac{7}{10}$

   d. $\frac{5}{21}$

238. Manuel has worked $6\frac{5}{8}$ hours of his regular eight-hour day. How many more hours must he work?

   a. $1\frac{1}{2}$ hours

   b. $1\frac{3}{8}$ hours

   c. $2\frac{1}{4}$ hours

   d. $1\frac{1}{4}$ hours

239. Irma has read $\frac{3}{5}$ of the novel assigned for her English class. The novel is 360 pages long. How many pages has she read?

   a. 216

   b. 72

   c. 300

   d. 98

240. Jerry rode his bike $7\frac{3}{4}$ miles on Monday, $5\frac{1}{5}$ miles on Tuesday, $6\frac{2}{5}$ miles on Wednesday, $7\frac{1}{2}$ miles on Thursday, $5\frac{1}{4}$ miles on Friday, and $6\frac{3}{5}$ miles on Saturday. How many total miles did he bike on those six days?

   a. 36 miles

   b. 38 miles

   c. $38\frac{7}{10}$ miles

   d. $38\frac{14}{15}$ miles

▶ **Set 16** (Answers begin on page 172.)

241. Meryl ordered a claw hammer, four drill bits, a work light, a large clamp, two screwdrivers, seven toggle bolts, 16 two-penny nails, three paintbrushes, and a 48-inch level from a mail order house. So far, she has received the hammer, three drill bits, the level, one screwdriver, the clamp, and all the two-penny nails. What fraction of her order has she received?

a. $\frac{1}{32}$

b. $\frac{16}{23}$

c. $\frac{23}{36}$

d. $\frac{36}{23}$

242. Lu makes $7.75 an hour. He worked $38\frac{1}{5}$ hours last week. How much money did he earn?

a. $592.10

b. $296.05

c. $775.00

d. $380.25

243. A lasagna recipe calls for $3\frac{1}{2}$ pounds of noodles. How many pounds of noodles are needed to make $\frac{1}{3}$ of a recipe?

a. 1 pound

b. $1\frac{1}{2}$ pounds

c. $\frac{5}{6}$ pound

d. $1\frac{1}{6}$ pounds

244. A lasagna recipe requires $1\frac{1}{2}$ pounds of cheese. Approximately how many lasagnas can be made from a $20\frac{1}{3}$-pound block of cheese?

a. $13\frac{1}{2}$

b. $20\frac{1}{3}$

c. $10\frac{1}{5}$

d. $25\frac{1}{4}$

245. For health reasons, Amir wants to drink eight glasses of water a day. He's already had six glasses. What fraction does Amir have left to drink?

a. $\frac{1}{8}$

b. $\frac{1}{6}$

c. $\frac{1}{4}$

d. $\frac{1}{3}$

246. Wendy is writing a test to give to her history class. She wants the test to include 40 multiple-choice questions and 60 short-answer questions. She has written 25 of the multiple-choice questions. What fraction of the total test has she written?

a. $\frac{1}{4}$

b. $\frac{5}{8}$

c. $\frac{2}{3}$

d. $\frac{5}{12}$

247. Malcolm's car gets $14\frac{1}{3}$ miles per gallon. It's $58\frac{1}{2}$ miles from his home to work. How many gallons does Malcolm's car use on the way to work?

a. $2\frac{9}{10}$ gallons

b. $3\frac{1}{16}$ gallons

c. $4\frac{7}{86}$ gallons

d. $5\frac{3}{8}$ gallons

248. Felicia needs 168 six-inch fabric squares to make a quilt top. She has 150 squares. What fraction of the total does she still need?

a. $\frac{25}{28}$

b. $\frac{9}{88}$

c. $\frac{25}{28}$

d. $\frac{3}{28}$

249. Roger wants to paint his living room ceiling red. His ceiling is $14\frac{1}{2}$ feet by $12\frac{1}{3}$ feet. One gallon of paint will cover 90 square feet. How many gallons of paint will he need?

   a. 1 gallon

   b. 2 gallons

   c. 3 gallons

   d. 4 gallons

250. The Garcias had $\frac{2}{5}$ of last night's meat loaf left over after dinner. Today, Uncle Jorge ate $\frac{1}{4}$ of these leftovers. How much of the original meat loaf is left?

   a. $\frac{3}{4}$

   b. $\frac{3}{10}$

   c. $\frac{3}{20}$

   d. $\frac{3}{5}$

251. Millie is a night security guard at the art museum. Each night, she is required to walk through each gallery once. The museum contains 52 galleries. This night, Millie has walked through 16 galleries. What fraction of the total galleries has she already visited?

   a. $\frac{4}{13}$

   b. $\frac{1}{16}$

   c. $\frac{5}{11}$

   d. $\frac{3}{14}$

252. Alan has been ill and worked only $\frac{3}{4}$ of his usual 40-hour week. He makes $12.35 an hour. How much has he earned this week?

   a. $247.00

   b. $308.75

   c. $370.50

   d. $432.25

253. A recipe calls for $\frac{1}{4}$ teaspoon of red pepper. How much red pepper would you need for half a recipe?

   a. $\frac{1}{10}$ teaspoon

   b. $\frac{1}{8}$ teaspoon

   c. $\frac{1}{6}$ teaspoon

   d. $\frac{1}{2}$ teaspoon

254. A recipe calls for $\frac{1}{4}$ teaspoon of red pepper. How much red pepper would you need for a double recipe?

   a. $\frac{1}{10}$ teaspoon

   b. $\frac{1}{8}$ teaspoon

   c. $\frac{1}{6}$ teaspoon

   d. $\frac{1}{2}$ teaspoon

255. Juana's lawn is 30 yards by 27 yards. Yesterday, Juana mowed $\frac{2}{3}$ of the lawn. How many square yards are left to be mowed today?

   a. 270 square yards

   b. 540 square yards

   c. 810 square yards

   d. 1,080 square yards

256. A thirty-minute time slot on a television network contains 24 minutes of comedy and six minutes of commercials. What fraction of the program time is devoted to commercials?

   a. $\frac{1}{6}$

   b. $\frac{1}{4}$

   c. $\frac{1}{3}$

   d. $\frac{1}{5}$

▶ **Set 17** (Answers begin on page 172.)

257. Pete's dog is on a special diet and has lost 15 pounds. If the dog has lost $\frac{1}{8}$ of its original weight, what was the original weight?
   a. 105 pounds
   b. 110 pounds
   c. 115 pounds
   d. 120 pounds

258. In a cashier contest, Ona packed $15\frac{1}{2}$ bags of groceries in three minutes. How many bags did she average per minute?
   a. $4\frac{1}{2}$
   b. 5
   c. $5\frac{1}{4}$
   d. $5\frac{1}{6}$

259. Cheryl lives $5\frac{1}{3}$ miles from where she works. When traveling to work, she walks to a bus stop $\frac{1}{4}$ of the way to catch a bus. How many miles away from her house is the bus stop?
   a. $5\frac{1}{3}$ miles
   b. $4\frac{1}{3}$ miles
   c. $2\frac{1}{3}$ miles
   d. $1\frac{1}{3}$ miles

260. At birth, Winston weighed $6\frac{1}{2}$ pounds. At one year of age, he weighed $23\frac{1}{8}$ pounds. How much weight, in pounds, did he gain?
   a. $16\frac{5}{8}$ pounds
   b. $16\frac{7}{8}$ pounds
   c. $17\frac{1}{6}$ pounds
   d. $17\frac{3}{4}$ pounds

261. Marcia wants to make muffins and needs $\frac{3}{4}$ cup sugar. She discovers, however, that she has only $\frac{2}{3}$ cup sugar. How much more sugar does she need?
   a. $\frac{1}{12}$ cup
   b. $\frac{1}{8}$ cup
   c. $\frac{1}{6}$ cup
   d. $\frac{1}{4}$ cup

262. How many inches are there in $3\frac{1}{3}$ yards?
   a. 126 inches
   b. 120 inches
   c. 160 inches
   d. 168 inches

263. Carlin's Candy Shop opened for business on Saturday with $22\frac{1}{4}$ pounds of fudge. During the day, they sold $17\frac{5}{8}$ pounds of fudge. How many pounds were left?
   a. $4\frac{1}{2}$ pounds
   b. $4\frac{5}{8}$ pounds
   c. $4\frac{7}{8}$ pounds
   d. $5\frac{3}{8}$ pounds

264. A child's swimming pool contains $20\frac{4}{5}$ gallons of water. If $3\frac{1}{3}$ gallons of water are splashed out of the pool while the children are playing, how many gallons of water are left?
   a. $16\frac{1}{15}$ gallons
   b. $16\frac{3}{5}$ gallons
   c. $17\frac{7}{15}$ gallons
   d. $17\frac{2}{3}$ gallons

265. During the month of May, $\frac{1}{6}$ of the buses in District A were in the garage for routine maintenance. In addition, $\frac{1}{8}$ of the buses were in for other repairs. If a total of 28 buses were in for maintenance and repairs, how many buses did District A have altogether?

   a. 80

   b. 84

   c. 91

   d. 96

266. On Monday, a kindergarten class uses $2\frac{1}{4}$ pounds of modeling clay the first hour, $4\frac{5}{8}$ pounds of modeling clay the second hour, and $\frac{1}{2}$ pound of modeling clay the third hour. How many pounds of clay does the class use during the three hours on Monday?

   a. $6\frac{3}{8}$ pounds

   b. $6\frac{7}{8}$ pounds

   c. $7\frac{1}{4}$ pounds

   d. $7\frac{3}{8}$ pounds

267. Three kittens weigh $2\frac{1}{3}$ pounds, $1\frac{5}{6}$ pounds, and $2\frac{2}{3}$ pounds. What is the total weight of the kittens?

   a. $6\frac{1}{3}$ pounds

   b. $6\frac{5}{6}$ pounds

   c. $7\frac{1}{6}$ pounds

   d. $7\frac{1}{3}$ pounds

268. If Rachel has worked a total of $26\frac{1}{4}$ hours so far this week, and has to work a total of $37\frac{1}{2}$ hours, how much longer does she have to work?

   a. $10\frac{1}{4}$ hours

   b. $11\frac{1}{4}$ hours

   c. $11\frac{3}{4}$ hours

   d. $13\frac{1}{2}$ hours

269. On Roy's daily jog, he travels a distance of $\frac{1}{2}$ mile to get to the track and $\frac{1}{2}$ mile to get home from the track. One lap around the track is $\frac{1}{4}$ mile. If Roy jogs five laps around the track, what is the total distance that he travels?

   a. $2\frac{1}{4}$ miles

   b. $2\frac{1}{2}$ miles

   c. 3 miles

   d. $3\frac{1}{4}$ miles

270. Suzy's pie recipe calls for $1\frac{1}{3}$ cups sugar. If she wants to add an additional $\frac{1}{3}$ cup to make the pie sweeter, how much sugar will she need in all?

   a. $1\frac{1}{9}$ cups

   b. $1\frac{1}{6}$ cups

   c. $1\frac{2}{9}$ cups

   d. $1\frac{2}{3}$ cups

271. Jonah hiked $7\frac{3}{8}$ miles on Friday, $6\frac{3}{10}$ miles on Saturday, and $5\frac{1}{5}$ miles on Sunday. How many miles did he hike in all?

   a. $18\frac{5}{8}$ miles

   b. $18\frac{7}{8}$ miles

   c. $19\frac{3}{5}$ miles

   d. $20\frac{1}{10}$ miles

272. Miguel owns $16\frac{3}{4}$ acres of land. If he buys another $2\frac{3}{5}$ acres, how many acres of land will he own in all?

   a. $18\frac{4}{5}$ acres

   b. $18\frac{9}{20}$ acres

   c. $19\frac{3}{10}$ acres

   d. $19\frac{7}{20}$ acres

▶ **Set 18** (Answers begin on page 173.)

273. It takes Paula 25 minutes to wash her car. If she has been washing her car for 15 minutes, what fraction of the job has she already completed?
- **a.** $\frac{3}{5}$
- **b.** $\frac{1}{2}$
- **c.** $\frac{4}{15}$
- **d.** $\frac{2}{5}$

274. It takes three firefighters $1\frac{2}{5}$ hours to clean their truck. At that same rate, how many hours would it take one firefighter to clean the same truck?
- **a.** $2\frac{4}{7}$ hours
- **b.** $3\frac{4}{5}$ hours
- **c.** $4\frac{1}{5}$ hours
- **d.** $4\frac{2}{5}$ hours

275. How many $5\frac{1}{4}$-ounce glasses can be completely filled from a $33\frac{1}{2}$-ounce container of juice?
- **a.** 4
- **b.** 5
- **c.** 6
- **d.** 7

276. If one pint is $\frac{1}{8}$ of a gallon, how many pints are there in $3\frac{1}{2}$ gallons of ice cream?
- **a.** $\frac{7}{16}$ pint
- **b.** $24\frac{1}{2}$ pints
- **c.** $26\frac{1}{16}$ pints
- **d.** 28 pints

277. Eric's walking speed is $2\frac{1}{2}$ miles per hour. If it takes Eric six minutes to walk from his home to the bus stop, how far is the bus stop from his home?
- **a.** $\frac{1}{8}$ mile
- **b.** $\frac{1}{4}$ mile
- **c.** $\frac{1}{2}$ mile
- **d.** 1 mile

278. The directions on an exam allow $2\frac{1}{2}$ hours to answer 50 questions. If you want to spend an equal amount of time on each of the 50 questions, about how much time should you allow for each one?
- **a.** 45 seconds
- **b.** $1\frac{1}{2}$ minutes
- **c.** 2 minutes
- **d.** 3 minutes

279. Which of these is equivalent to 35° C?
($F = \frac{9}{5}C + 32$)
- **a.** 105° F
- **b.** 95° F
- **c.** 63° F
- **d.** 19° F

280. A firefighter checks the gauge on a cylinder that normally contains 45 cubic feet of air and finds that the cylinder has only 10 cubic feet of air. The gauge indicates that the cylinder is
- **a.** $\frac{1}{4}$ full.
- **b.** $\frac{2}{9}$ full.
- **c.** $\frac{1}{3}$ full.
- **d.** $\frac{4}{5}$ full.

281. If the diameter of a metal spool is 3.5 feet, how many times will a 53-foot hose wrap completely around it? ($C = \pi d$; $\pi = \frac{22}{7}$)

   a. 2

   b. 3

   c. 4

   d. 5

282. The high temperature in a certain city was 113° F. At about what temperature Celsius was this temperature? $C = \frac{5}{9}(F - 32)$

   a. 45° C

   b. 45.5° C

   c. 51° C

   d. 81.5° C

283. Tank A, when full, holds 555 gallons of water. Tank B, when full, holds 680 gallons of water. If Tank A is only $\frac{2}{3}$ full and Tank B is only $\frac{2}{5}$ full, how many more gallons of water are needed to fill both tanks to capacity?

   a. 319 gallons

   b. 593 gallons

   c. 642 gallons

   d. 658 gallons

284. At a party there are three large pizzas. Each pizza has been cut into nine equal pieces. Eight-ninths of the first pizza have been eaten; $\frac{2}{3}$ of the second pizza have been eaten; $\frac{7}{9}$ of the third pizza have been eaten. What fraction of the three pizzas is left?

   a. $\frac{2}{9}$

   b. $\frac{2}{7}$

   c. $\frac{1}{3}$

   d. $\frac{1}{6}$

285. If it takes four firefighters 1 hour 45 minutes to perform a particular job, how long would it take one firefighter working at the same rate to perform the same task alone?

   a. $4\frac{1}{2}$ hours

   b. 5 hours

   c. 7 hours

   d. $7\frac{1}{2}$ hours

286. Joel can change one lightbulb in $\frac{5}{6}$ minute. Working at that same rate, how many minutes would it take him to change five lightbulbs?

   a. $4\frac{1}{6}$ minutes

   b. $4\frac{1}{3}$ minutes

   c. $4\frac{2}{3}$ minutes

   d. $5\frac{1}{6}$ minutes

287. Ryan has two bags of jelly beans. One weighs $10\frac{1}{4}$ ounces; the other weighs $9\frac{1}{8}$ ounces. If Ryan puts the two bags together and then divides all of the jelly beans into five equal parts to give to his friends, how many ounces will each friend get?

   a. $3\frac{3}{4}$ ounces

   b. $3\frac{7}{8}$ ounces

   c. 4 ounces

   d. $4\frac{1}{4}$ ounces

288. At a certain school, half the students are female and $\frac{1}{12}$ of the students are from outside the state. What proportion of the students would you expect to be females from outside the state?

   a. $\frac{1}{12}$

   b. $\frac{1}{24}$

   c. $\frac{1}{6}$

   d. $\frac{1}{3}$

▶ Set 19 (Answers begin on page 174.)

289. How many minutes are in $7\frac{1}{6}$ hours?

   a. 258 minutes

   b. 430 minutes

   c. 2,580 minutes

   d. 4,300 minutes

290. One lap on a particular outdoor track measures $\frac{1}{4}$ mile around. To run a total of $3\frac{1}{2}$ miles, how many complete laps must a person run?

   a. 14

   b. 18

   c. 7

   d. 10

291. During the month of June, Bus #B-461 used the following amounts of oil:

June 1: $3\frac{1}{2}$ quarts

June 19: $2\frac{3}{4}$ quarts

June 30: 4 quarts

What is the total number of quarts used in June?

   a. $9\frac{3}{4}$ quarts

   b. 10 quarts

   c. $10\frac{1}{4}$ quarts

   d. $10\frac{1}{2}$ quarts

292. How many ounces are in $9\frac{1}{2}$ pounds?

   a. 192 ounces

   b. 182 ounces

   c. 152 ounces

   d. 132 ounces

293. Crystal's yearly income is $25,000, and the cost of her rent for the year is $7,500. What fraction of her yearly income does she spend on rent?

   a. $\frac{1}{4}$

   b. $\frac{3}{10}$

   c. $\frac{2}{5}$

   d. $\frac{2}{7}$

294. Julie counts the cars passing her house and finds that two of every five cars are foreign. If she counts for one hour, and 60 cars pass, how many of them are likely to be domestic?

   a. 12

   b. 24

   c. 30

   d. 36

295. A recipe calls for $1\frac{1}{4}$ cups flour. If Larry wants to make $2\frac{1}{2}$ times the recipe, how much flour does he need?

   a. $2\frac{3}{4}$ cups

   b. $3\frac{1}{8}$ cups

   c. $3\frac{1}{4}$ cups

   d. $3\frac{5}{8}$ cups

296. An auditorium that holds 350 people currently has 150 seated in it. What part of the auditorium is full?

   a. $\frac{1}{4}$

   b. $\frac{1}{3}$

   c. $\frac{3}{7}$

   d. $\frac{3}{5}$

297. Third-grade student Stephanie goes to the school nurse's office, where her temperature is found to be 98° Fahrenheit. What is her temperature in degrees Celsius? $C = \frac{5}{9}(F - 32)$
- **a.** 35.8° C
- **b.** 36.7° C
- **c.** 37.6° C
- **d.** 31.1° C

298. The temperature recorded at 8 A.M. is 30° C. What is the equivalent of this temperature in degrees Fahrenheit? $F = \frac{9}{5}C + 32$
- **a.** 59° F
- **b.** 62° F
- **c.** 86° F
- **d.** 95° F

299. If Tory donates $210 to charitable organizations each year and $\frac{1}{3}$ of that amount goes to the local crisis center, how much of her yearly donation does the crisis center get?
- **a.** $33.00
- **b.** $45.50
- **c.** $60.33
- **d.** $70.00

300. A construction job calls for $2\frac{5}{6}$ tons of sand. Four trucks, each filled with $\frac{3}{4}$ tons of sand, arrive on the job. Is there enough sand, or is there too much sand for the job?
- **a.** There is not enough sand; $\frac{1}{6}$ ton more is needed.
- **b.** There is not enough sand; $\frac{1}{3}$ ton more is needed.
- **c.** There is $\frac{1}{3}$ ton more sand than is needed.
- **d.** There is $\frac{1}{6}$ ton more sand than is needed.

301. A safety box has three layers of metal, each with a different width. If one layer is $\frac{1}{8}$-inch thick, a second layer is $\frac{1}{6}$-inch thick, and the total thickness is $\frac{3}{4}$-inch thick, what is the width of the third layer?
- **a.** $\frac{5}{12}$ inch
- **b.** $\frac{11}{24}$ inch
- **c.** $\frac{7}{18}$ inch
- **d.** $\frac{1}{2}$ inch

Answer questions 302 and 303 using the following list of ingredients needed to make 16 brownies.

Deluxe Brownies
$\frac{2}{3}$ cup butter
5 squares (1 ounce each) unsweetened chocolate
$1\frac{1}{2}$ cups sugar
2 teaspoons vanilla
2 eggs
1 cup flour

302. How many brownies can be made if the baker increases the recipe to include one cup of butter?
- **a.** 12 brownies
- **b.** 16 brownies
- **c.** 24 brownies
- **d.** 28 brownies

303. How much sugar is needed in a recipe that makes eight brownies?
- **a.** $\frac{3}{4}$ cup
- **b.** 3 cups
- **c.** $\frac{2}{3}$ cup
- **d.** $\frac{5}{8}$ cup

304. George cuts his birthday cake into 10 equal pieces. If six people eat a piece of George's cake, what fraction of the cake is left?

a. $\frac{3}{5}$

b. $\frac{3}{10}$

c. $\frac{2}{5}$

d. $\frac{5}{6}$

▶ **Set 20** (Answers begin on page 175.)

305. A certain congressional district has about 490,000 people living in it. The largest city in the area has 98,000 citizens. Which most accurately portrays the portion of the population made up by the city in the district?

   a. $\frac{1}{5}$

   b. $\frac{1}{4}$

   c. $\frac{2}{9}$

   d. $\frac{3}{4}$

306. A bag of jelly beans contains eight black beans, 10 green beans, three yellow beans, and nine orange beans. What is the probability of selecting either a yellow or an orange bean?

   a. $\frac{1}{10}$

   b. $\frac{2}{5}$

   c. $\frac{4}{15}$

   d. $\frac{3}{10}$

307. Each piece of straight track for Ty's electric train set is $6\frac{1}{2}$ inches long. If five pieces of this track are laid end to end, how long will the track be?

   a. $30\frac{1}{2}$ inches

   b. 32 inches

   c. $32\frac{1}{2}$ inches

   d. $32\frac{5}{8}$ inches

308. How many $\frac{1}{4}$-pound hamburgers can be made from six pounds of ground beef?

   a. 18 hamburgers

   b. $20\frac{1}{2}$ hamburgers

   c. 24 hamburgers

   d. $26\frac{1}{4}$ hamburgers

309. Bart's eight-ounce glass is $\frac{4}{5}$ full of water. How many ounces of water does he have?

   a. $4\frac{5}{8}$ ounces

   b. 5 ounces

   c. 6 ounces

   d. $6\frac{2}{5}$ ounces

310. Barbara can walk $3\frac{1}{4}$ miles in one hour. At that rate, how many miles will she walk in $1\frac{2}{3}$ hours?

   a. $4\frac{5}{8}$ miles

   b. $4\frac{11}{12}$ miles

   c. $5\frac{5}{12}$ miles

   d. 6 miles

311. Three friends evenly split $1\frac{1}{8}$ pounds of peanuts. How many pounds will each person get?

   a. $\frac{1}{4}$ pound

   b. $\frac{3}{8}$ pound

   c. $\frac{1}{2}$ pound

   d. $\frac{5}{8}$ pound

312. Iris lives $2\frac{1}{2}$ miles due east of the Sunnydale Mall, and Raoul lives $4\frac{1}{2}$ miles due west of the QuikMart, which is $1\frac{1}{2}$ miles due west of the Sunnydale Mall. How far does Iris live from Raoul?

   a. 8 miles

   b. 8.5 miles

   c. 9 miles

   d. 9.5 miles

313. Ingrid's kitchen is $9\frac{3}{4}$ feet long and $8\frac{1}{3}$ feet wide. How many square feet of tile does she need to tile the floor?

a. $81\frac{1}{4}$ square feet

b. $72\frac{1}{4}$ square feet

c. $71\frac{1}{2}$ square feet

d. $82\frac{1}{2}$ square feet

314. How many inches are in $4\frac{1}{2}$ feet?

a. 48 inches

b. 54 inches

c. 66 inches

d. 70 inches

315. During the winter, Lucas missed $7\frac{1}{2}$ days of kindergarten due to colds, while Brunhilda missed only $4\frac{1}{4}$ days. How many fewer days did Brunhilda miss than Lucas?

a. $3\frac{1}{4}$

b. $3\frac{1}{2}$

c. $3\frac{5}{6}$

d. 4

316. To reach his tree house, Raymond has to climb $9\frac{1}{3}$ feet up a rope ladder, then $8\frac{5}{6}$ feet up the tree trunk. How far does Raymond have to climb altogether?

a. $17\frac{7}{12}$ feet

b. $17\frac{1}{6}$ feet

c. $18\frac{1}{6}$ feet

d. $18\frac{1}{2}$ feet

317. Ralph's newborn triplets weigh $4\frac{3}{8}$ pounds, $3\frac{5}{6}$ pounds, and $4\frac{7}{8}$ pounds. Harvey's newborn twins weigh $7\frac{2}{6}$ pounds and $9\frac{3}{10}$ pounds. Whose babies weigh the most and by how much?

a. Ralph's triplets by $3\frac{1}{2}$ pounds

b. Ralph's triplets by $2\frac{1}{4}$ pounds

c. Harvey's twins by $1\frac{2}{3}$ pounds

d. Harvey's twins by $3\frac{11}{20}$ pounds

318. Sofia bought a pound of vegetables and used $\frac{3}{8}$ of it to make a salad. How many ounces of vegetables are left after she makes the salad?

a. 4 ounces

b. 6 ounces

c. 8 ounces

d. 10 ounces

319. Dani spent $6,300 on a used car. She paid $630 as a down payment. What fraction of the original cost was the down payment?

a. $\frac{1}{10}$

b. $\frac{1}{18}$

c. $\frac{1}{20}$

d. $\frac{1}{40}$

320. During an eight-hour workday, Bob spends two hours on the phone. What fraction of the day does he spend on the phone?

a. $\frac{1}{5}$

b. $\frac{1}{3}$

c. $\frac{1}{4}$

d. $\frac{1}{8}$

3 ▶ Decimals

These 10 sets of problems will familiarize you with arithmetic operations involving decimals (which are a really special kind of fraction). You use decimals every day, in dealing with money, for example. Units of measurement, such as populations, kilometers, inches, or miles, are also often expressed in decimals. In this section you will get practice in working with *mixed decimals*, or numbers that have digits on both sides of a decimal point, and the important tool of *rounding*, the method for estimating decimals.

▶ **Set 21** (Answers begin on page 177.)

321. 56.73647 rounded to the nearest hundredth is equal to
- **a.** 100
- **b.** 57
- **c.** 56.7
- **d.** 56.74

322. Which number sentence is true?
- **a.** 0.43 < 0.043
- **b.** 0.0043 > 0.43
- **c.** 0.00043 > 0.043
- **d.** 0.043 > 0.0043

323. 78.09 + 19.367 =
- **a.** 58.723
- **b.** 87.357
- **c.** 97.457
- **d.** 271.76

324. 3.419 − 0.7 =
- **a.** 34.12
- **b.** 0.2719
- **c.** 2.719
- **d.** 0.3412

325. 2.9 ÷ 0.8758 =
- **a.** 3.31
- **b.** 0.331
- **c.** 0.302
- **d.** 0.0302

326. 195.6 ÷ 7.2, rounded to the nearest hundredth, is equal to
- **a.** 271.67
- **b.** 27.17
- **c.** 27.16
- **d.** 2.717

327. 515 − 4.2 =
- **a.** 51.08
- **b.** 5.108
- **c.** 510.8
- **d.** 5,108

328. 7.14 × 7 =
- **a.** 49.98
- **b.** 499.8
- **c.** 4.98
- **d.** 49

329. 9.3 − 8.132 =
- **a.** 1,168
- **b.** 1.168
- **c.** 11.68
- **d.** 1.68

330. 824 × 0.18 =
- **a.** 0.14832
- **b.** 148.32
- **c.** 14.832
- **d.** 14,832

331. 0.34 × 0.56 =
- **a.** 19.04
- **b.** 1.904
- **c.** 0.194
- **d.** 0.1904

332. What is five and four hundredths written as a decimal?
- **a.** 0.54
- **b.** 0.054
- **c.** 5.4
- **d.** 5.04

333. In the following decimal, which digit is in the hundredths place: 0.2153

a. 2
b. 1
c. 5
d. 3

334. What is 0.2275 rounded to the nearest tenth?

a. 0.2
b. 0.3
c. 0.22
d. 0.5

335. Which of these has an 8 in the thousandths place?

a. 1.68
b. 1.0068
c. 1.0086
d. 1.8006

336. $6.75 \div 6.25 =$

a. 1
b. 1.08
c. 1.8
d. 12

▶ **Set 22** (Answers begin on page 177.)

337. $0.321 + 6.5 + 64 =$
- a. 70.821
- b. 391.5
- c. 1.611
- d. 708.21

338. $0.42 \times 0.09 =$
- a. 3.78
- b. 37.8
- c. 0.0378
- d. 0.378

339. $134.81 \div 34 =$
- a. 3.965
- b. 0.3965
- c. 39.65
- d. 3.0965

340. $26.907 + 234.76 =$
- a. 0.26167
- b. 261.667
- c. 26.1667
- d. 2,616.67

341. $0.0224 + 0.0569 =$
- a. 0.0793
- b. 0.793
- c. 0.7
- d. 0.739

342. $4.2 - 2.37 =$
- a. 2.37
- b. 1.95
- c. 6.57
- d. 1.83

343. What is another way to write 4.32×100^2?
- a. 432
- b. 4,320
- c. 43,200
- d. 432,000

344. $0.476 \div 7 =$
- a. 6.8
- b. 0.68
- c. 0.068
- d. 0.0068

345. $489.05 \times 0.25 =$
- a. 0.12625
- b. 12,625
- c. 1,222.625
- d. 122.2625

346. What is another way to write 7.25×10^3?
- a. 72.5
- b. 725
- c. 7,250
- d. 72,500

347. $(4.1 \times 10^{-2})(3.8 \times 10^4) =$
- a. 1.558×10^{-8}
- b. 15.58×10^{-2}
- c. 1.558×10^2
- d. 1.558×10^3

348. $\frac{6.5 \times 10^{-6}}{3.25 \times 10^{-3}} =$
- a. 2×10^{-9}
- b. 2×10^{-3}
- c. 2×10^2
- d. 2×10^3

349. What is the product of 16 and 0.023?
   a. 0.368
   b. 0.0368
   c. 3.68
   d. 0.36

350. 0.75 + 0.518 =
   a. 12.68
   b. 0.01268
   c. 0.1268
   d. 1.268

351. Which of the following numbers is NOT between −0.06 and 1.06?
   a. 0
   b. 0.06
   c. −0.16
   d. −0.016

▶ **Set 23** (Answers begin on page 178.)

352. Which of the following is the word form of the decimal 0.08?
- **a.** eight hundredths
- **b.** eight tenths
- **c.** eight thousandths
- **d.** eight ten-thousandths

353. Which of the following decimals has the greatest value?
- **a.** 8.241
- **b.** 8.0241
- **c.** 8.2
- **d.** 8.2041

354. Which of the following decimals has the least value?
- **a.** 0.97
- **b.** 0.0907
- **c.** 0.097
- **d.** 0.0097

355. What is 43.0089 rounded to the nearest hundredth?
- **a.** 43.008
- **b.** 43.01
- **c.** 43.008
- **d.** 43.1

356. What is the sum of $13.008 + 52 + 0.69$ rounded to the nearest tenth?
- **a.** 65
- **b.** 65.7
- **c.** 65.077
- **d.** 65.07

357. $18.7772 + 23 + 0.8789 =$
- **a.** 42.6561
- **b.** 42.0656
- **c.** 43.6165
- **d.** 43.6561

358. $4 - 0.93 - 0.2 =$
- **a.** 2.0087
- **b.** 28.7
- **c.** 2.87
- **d.** 2.087

359. Yuri and Catherine begin driving at the same time but in opposite directions. If Yuri drives 60 miles per hour, and Catherine drives 70 miles per hour, approximately how long will it be before they are 325.75 miles apart?
- **a.** 2 hours
- **b.** 2.25 hours
- **c.** 2.5 hours
- **d.** 2.75 hours

360. A garden store offers $0.75 off all pumpkins after Halloween. If a certain pumpkin is priced at $3.20 after Halloween, what was the original price?
- **a.** $2.45
- **b.** $3.75
- **c.** $3.95
- **d.** $4.05

361. A bottle of apple juice contains 1.42 liters; a bottle of grape juice contains 1.89 liters. How many total liters of juice are there in the two bottles?
- **a.** 0.47 liters
- **b.** 2.31 liters
- **c.** 3.21 liters
- **d.** 3.31 liters

362. Over the last few years, the average number of dogs in the neighborhood has dropped from 17.8 to 14.33. What is the decrease in the number of dogs in the neighborhood?

a. 3.47

b. 2.66

c. 3.15

d. 2.75

363. On Monday, Freida slept 6.45 hours; on Tuesday, 7.32; on Wednesday, 5.1; on Thursday, 6.7; and on Friday, she slept 8.9 hours. How many hours of sleep did she get over the five days?

a. 40.34 hours

b. 34.47 hours

c. 36.78 hours

d. 42.67 hours

364. Last week, Felicity had $67.98 saved from babysitting. She made another $15.75 babysitting this week and spent $27.58 on CDs. How much money does she have now?

a. $71.55

b. $24.65

c. $111.31

d. $56.15

365. Roommates Bob and Ted agreed to wallpaper and carpet the living room and replace the sofa. The wallpaper costs $103.84, the carpet costs $598.15, and the new sofa costs $768.56. Bob agrees to pay for the carpet and wallpaper, and Ted agrees to pay for the sofa. How much more money will Ted spend than Bob?

a. $68.56

b. $76.55

c. $66.57

d. $72.19

366. Reams of copy paper cost $11.39 for five cases. How much would 100 cases cost?

a. $56.95

b. $227.80

c. $1,139.00

d. $68.34

367. A bartender earned a total of $87 in tips over six days. What is the average amount of total tips earned per day?

a. $15.50

b. $14.00

c. $14.50

d. $15.00

368. Eight coworkers decide to split a dinner bill evenly, including tip. Each coworker paid $23.65. What was the total cost of the bill, including tip?

a. $188.25

b. $189.20

c. $189.00

d. $188.65

► **Set 24** (Answers begin on page 179.)

369. Samuel paid $5.96 for four pounds of cookies. How much do the cookies cost per pound?
- **a.** $1.96
- **b.** $2.33
- **c.** $1.49
- **d.** $2.15

370. One inch equals 2.54 centimeters. How many centimeters are there in a foot?
- **a.** 30.48 centimeters
- **b.** 32.08 centimeters
- **c.** 31.79 centimeters
- **d.** 29.15 centimeters

371. Melinda and Joaquin can restock an aisle at the supermarket in one hour working together. Melinda can restock an aisle in 1.5 hours working alone, and it takes Joaquin two hours to restock an aisle. If they work together for two hours, and then work separately for another two hours, how many aisles will they have completed?
- **a.** 5
- **b.** 4.5
- **c.** 4.33
- **d.** 3.5

372. Patrick earns only $\frac{1}{3}$ of what Robin does. If Robin makes $21 per hour, how much does Patrick earn in a typical eight-hour workday?
- **a.** $56
- **b.** $112
- **c.** $168
- **d.** $7

373. Reva earns $10 an hour for walking the neighbor's dog. Today she can walk the dog for only 45 minutes. How much will Reva make today?
- **a.** $6.25
- **b.** $7.50
- **c.** $7.75
- **d.** $8.00

374. A greyhound, Zelda, can run 35.25 miles an hour, while a cat, Spot, can run only $\frac{1}{4}$ that fast. How many miles per hour can Spot run?
- **a.** 8.25 miles per hour
- **b.** 8.77 miles per hour
- **c.** 8.81 miles per hour
- **d.** 9.11 miles per hour

375. A 600-page book is 1.5 inches thick. What is the thickness of each page?
- **a.** 0.0010 inch
- **b.** 0.0030 inch
- **c.** 0.0025 inch
- **d.** 0.0600 inch

376. Michael walks to school. He leaves each morning at 7:32 A.M. and arrives at school 15 minutes later. If he travels at a steady rate of 4.5 miles per hour, what is the distance between his home and the school? (*Distance = rate × time*)
- **a.** 1.1 miles
- **b.** 1.125 miles
- **c.** 1.5 miles
- **d.** 2.5 miles

377. For every dollar Kyra saves, her employer contributes a dime to her savings, with a maximum employer contribution of $10 per month. If Kyra saves $60 in January, $130 in March, and $70 in April, how much will she have in savings at the end of that time?
 a. $270
 b. $283
 c. $286
 d. $290

378. Quentin was shopping for a new washing machine. The one he wanted to buy cost $428.98. The salesperson informed him that the same machine would be on sale the following week for $399.99. How much money would Quentin save by waiting until the washing machine went on sale?
 a. $28.99
 b. $29.01
 c. $39.09
 d. $128.99

379. If a bus weighs 2.5 tons, how many pounds does it weigh? (one ton = 2,000 pounds)
 a. 800 pounds
 b. 4,500 pounds
 c. 5,000 pounds
 d. 5,500 pounds

380. If Serena burns about 304.15 calories while walking fast on her treadmill for 38.5 minutes, about how many calories does she burn per minute?
 a. 7.8
 b. 7.09
 c. 7.9
 d. 8.02

381. A truck is carrying 1,000 television sets; each set weighs 21.48 pounds. What is the total weight, in pounds, of the entire load?
 a. 214.8 pounds
 b. 2,148 pounds
 c. 21,480 pounds
 d. 214,800 pounds

382. Luis is mailing two packages. One weighs 12.9 pounds, and the other weighs half as much. What is the total weight, in pounds, of the two packages?
 a. 6.45 pounds
 b. 12.8 pounds
 c. 18.5 pounds
 d. 19.35 pounds

383. If it takes Danielle 22.4 minutes to walk 1.25 miles, how many minutes will it take her to walk one mile?
 a. 17.92 minutes
 b. 18 minutes
 c. 19.9 minutes
 d. 21.15 minutes

384. Mark's temperature at 9:00 A.M. was 97.2° F. At 4:00 P.M., his temperature was 99° F. By how many degrees did his temperature rise?
 a. 0.8°
 b. 1.8°
 c. 2.2°
 d. 2.8°

▶ Set 25 (Answers begin on page 180.)

385. Jason had $40 in his wallet. He bought gasoline for $12.90, a pack of gum for $0.45, and a candy bar for $0.88. How much money did he have left?
- **a.** $14.23
- **b.** $25.77
- **c.** $25.67
- **d.** $26.77

386. The price of cheddar cheese is $2.12 per pound. The price of Monterey Jack cheese is $2.34 per pound. If Harrison buys 1.5 pounds of cheddar and one pound of Monterey Jack, how much will he spend in all?
- **a.** $3.18
- **b.** $4.46
- **c.** $5.41
- **d.** $5.52

387. From a 100-foot ball of string, Randy cuts three pieces of the following lengths: 5.8 feet, 3.2 feet, and 4.4 feet. How many feet of string are left?
- **a.** 66.0 feet
- **b.** 86.6 feet
- **c.** 87.6 feet
- **d.** 97.6 feet

388. Fifteen-ounce cans of clam chowder sell at three for $2. How much does one can cost, rounded to the nearest cent?
- **a.** $0.60
- **b.** $0.66
- **c.** $0.67
- **d.** $0.70

389. For the month of July, Victoria purchased the following amounts of gasoline for her car: 9.4 gallons, 18.9 gallons, and 22.7 gallons. How many gallons did she purchase in all?
- **a.** 51 gallons
- **b.** 51.9 gallons
- **c.** 52 gallons
- **d.** 61 gallons

390. Manny works Monday through Friday each week. His bus fare to and from work is $1.10 each way. How much does Manny spend on bus fare each week?
- **a.** $10.10
- **b.** $11.00
- **c.** $11.10
- **d.** $11.20

391. If one centimeter equals 0.39 inches, about how many centimeters are there in 0.75 inches?
- **a.** 0.2925 centimeter
- **b.** 1.923 centimeters
- **c.** 0.52 centimeter
- **d.** 1.75 centimeters

392. 5.133 multipled by 10^{-6} is equal to
- **a.** 0.0005133
- **b.** 0.00005133
- **c.** 0.000005133
- **d.** 0.0000005133

393. On a business trip, Felicia went out to lunch. The shrimp cocktail cost $5.95, the blackened swordfish with grilled vegetables cost $11.70, the cherry cheesecake cost $4.79, and the coffee was $1.52. What was Felicia's total bill not including tax and tip?

a. $23.96

b. $26.93

c. $29.63

d. $32.99

394. Hal wants to buy a used car to take to college. The car costs $4,999.95. For graduation, he receives gifts of $200.00, $157.75, and $80.50. His little brother gave him $1.73, and he saved $4,332.58 from his summer job. How much more money does he need?

a. $272.93

b. $227.39

c. $722.93

d. $772.39

395. A football team must move the ball 10 yards in four plays in order to keep possession of the ball. The home team has just run a play in which they gained 2.75 yards. How many yards remain in order to keep possession of the ball?

a. 6.50 yards

b. 6.75 yards

c. 7.25 yards

d. 8.25 yards

396. Heather wants to build a deck that is 8.245 feet by 9.2 feet. How many square feet will the deck be? (*Area = length × width*)

a. 34.9280 square feet

b. 82.4520 square feet

c. 17.4450 square feet

d. 75.8540 square feet

397. Rene's car used 23.92 gallons of fuel on a 517.6-mile trip. How many miles per gallon did the car get, rounded to the nearest hundredth?

a. 18.46 miles per gallon

b. 21.64 miles per gallon

c. 26.14 miles per gallon

d. 29.61 miles per gallon

398. Faron ran a 200-meter race in 23.7 seconds. Gene ran it in 22.59 seconds. How many fewer seconds did it take Gene than Faron?

a. 2.17 seconds

b. 0.97 second

c. 1.01 seconds

d. 1.11 seconds

399. Mabel buys 2.756 pounds of sliced turkey, 3.2 pounds of roast beef, and 5.59 pounds of bologna. Approximately how many total pounds of meat did Mabel buy?

a. 11.5 pounds

b. 12.75 pounds

c. 10.03 pounds

d. 13.4 pounds

400. Kathleen types about 41.46 words per minute. At this rate, about how many words will she type in eight minutes?

a. 5.18

b. 33.46

c. 330.88

d. 331.68

▶ **Set 26** (Answers begin on page 180.)

401. Ingrid has 7.5 pounds of candy for trick-or-treaters. She gives a vampire 0.25 pounds, a fairy princess 0.53 pounds, and a horse 1.16 pounds. She eats 0.12 pounds. How much candy is left?

   a. 5.44 pounds

   b. 4.55 pounds

   c. 5.3 pounds

   d. 4.7 pounds

402. Ophelia drives from home to the grocery store, which is 6.2 miles. Then she goes to the video store, which is 3.4 more miles. Next, she goes to the bakery, which is 0.82 more miles. Then she drives the 5.9 miles home. How many miles total did she drive?

   a. 12.91 miles

   b. 13.6 miles

   c. 16.32 miles

   d. 18.7 miles

403. Ken wants to make slip covers for his dining room chairs. Each chair requires 4.75 yards of fabric. Ken has 20.34 yards of fabric. How many chairs can he cover?

   a. 3

   b. 4

   c. 5

   d. 6

404. On Monday, Luna had $792.78 in her checking account. On Tuesday, she deposited her $1,252.60 paycheck. On Wednesday, she paid her rent, $650. On Thursday, she paid her electric, cable, and phone bills, which were $79.35, $54.23, and $109.56, respectively. How much money is left in Luna's account?

   a. $965.73

   b. $1,348.90

   c. $893.14

   d. $1,152.24

405. Mya has a rectangular frame with an opening that is 11.25 inches by 8.75 inches. What is the area, rounded to the nearest hundredth, of the opening?

   a. 20 square inches

   b. 92.81 square inches

   c. 97.43 square inches

   d. 98.44 square inches

406. Michael's favorite cake recipe calls for 0.75 pounds of flour; he has a five-pound bag. He wants to make several cakes for the school bake sale. How many cakes can he make?

   a. 5

   b. 6

   c. 7

   d. 8

407. Hannah walks to work every day, sometimes running errands on the way. On Monday, she walked 0.75 mile; Tuesday, 1.2 miles; Wednesday, 1.68 miles; Thursday, 0.75 mile; and on Friday, she rode with Earl. On the days she walked, what was the average distance Hannah walked each day?
- **a.** 2.78 miles
- **b.** 1.231 miles
- **c.** 0.75 mile
- **d.** 1.095 miles

408. Seven people are at the beach for a clambake. They have dug 12.6 pounds of clams. They each eat the following amounts of clams: 0.34 pound, 1.6 pounds, 0.7 pound, 1.265 pounds, 0.83 pound, 1.43 pounds, 0.49 pound. How many pounds of clams are left?
- **a.** 7.892 pounds
- **b.** 4.56 pounds
- **c.** 5.945 pounds
- **d.** 6.655 pounds

409. A certain professional baseball player makes $2.4 million a year. A certain professional football player makes $1.025 million a year. How much less per year is the football player making than the baseball player?
- **a.** $1.375 million
- **b.** $1.15 million
- **c.** $1.46 million
- **d.** $2.46 million

410. A carpet costs $2.89 per square foot. How much carpet, rounded to the nearest square foot, could be bought with $76?
- **a.** 25 square feet
- **b.** 26 square feet
- **c.** 27 square feet
- **d.** 38 square feet

411. Philip has worked 34.75 hours of his usual 39.5-hour week. How many hours does he have left to work?
- **a.** 5.25 hours
- **b.** 4.75 hours
- **c.** 4.00 hours
- **d.** 3.75 hours

412. Gail has to water a 0.25-acre garden on Monday, a 1.02-acre garden on Tuesday, and a 0.36-acre garden on Wednesday. How many acres total does she have to water in the three days?
- **a.** 1.63 acres
- **b.** 1.53 acres
- **c.** 1.38 acres
- **d.** 1.27 acres

413. Lilly is going to recarpet her living room. The dimensions of the room are 15.6 feet by 27.75 feet. How many square feet of carpet will she need?
- **a.** 315.8 square feet
- **b.** 409.5 square feet
- **c.** 329.25 square feet
- **d.** 432.9 square feet

414. A movie is scheduled for two hours. The theater advertisements are 3.8 minutes long. There are two previews; one is 4.6 minutes long, and the other is 2.9 minutes long. The rest of time is devoted to the feature. How long is the feature film?

a. 108.7 minutes

b. 97.5 minutes

c. 118.98 minutes

d. 94.321 minutes

415. Tommy is making hats for the kids in the neighborhood. He needs 0.65 yard of fabric for each hat; he has 10 yards of fabric. How many hats can he make?

a. 13

b. 14

c. 15

d. 16

416. Over the weekend, Maggie watched 11.78 hours of television, Jenny watched 6.9 hours, Chas watched 7 hours, and Manny watched 2.45 hours. How much television did they watch over the weekend?

a. 26.786 hours

b. 28.13 hours

c. 30.79 hours

d. 32.85 hours

▶ **Set 27** (Answers begin on page 181.)

417. Carol wants to enclose a rectangular area in her backyard for her children's swing set. This section of the yard measures 16.25 feet by 20.25 feet. How many feet of fencing will she need to enclose this section?
   a. 36.5 feet
   b. 52.75 feet
   c. 73 feet
   d. 329.06 feet

418. In the new television season, an average of 7.9 million people watched one network; an average of 8.6 million people watched another. How many more viewers did the second network average than the first?
   a. 0.5 million
   b. 0.6 million
   c. 0.7 million
   d. 0.8 million

419. A writer makes $1.13 per book sold. How much will she make when 100 books have been sold?
   a. $11.30
   b. $113.00
   c. $1,130.00
   d. $11,300.00

420. Fred walks 0.75 mile to school; Ramona walks 1.3 miles; Xena walks 2.8 miles; and Paul walks 0.54 mile. What is the total distance the four walk to school?
   a. 4.13 miles
   b. 5.63 miles
   c. 4.78 miles
   d. 5.39 miles

421. Jane is driving 46.75 miles per hour. How far will she go in 15 minutes?
   a. 14.78 miles
   b. 11.6875 miles
   c. 12.543 miles
   d. 10.7865 miles

422. After stopping at a rest stop, Jane continues to drive at 46.75 miles per hour. How far will she go in 3.80 hours?
   a. 177.65 miles
   b. 213.46 miles
   c. 143.78 miles
   d. 222.98 miles

423. Bob notices that the ratio of boys to the total students in his class is 3:4. If there are 28 students in his class, how many of them are boys?
   a. 7
   b. 14
   c. 21
   d. 24

424. District C spends about $4,446.00 on diesel fuel each week. If the cost of diesel fuel to the district is about $1.17 per gallon, about how many gallons of diesel fuel does the district use in one week?
   a. 3,800 gallons
   b. 3,810 gallons
   c. 3,972 gallons
   d. 5,202 gallons

425. If one inch equals 2.54 centimeters, how many inches are there in 20.32 centimeters?
   a. 7.2 inches
   b. 8 inches
   c. 9 inches
   d. 10.2 inches

426. Kendra earns $12.50 an hour. When she works more than eight hours in one day, she earns $1\frac{1}{2}$ times her regular hourly wage. If she earns $137.50 for one day's work, how many hours did she work that day?
  a. 8.5 hours
  b. 9 hours
  c. 10 hours
  d. 11 hours

427. If a car travels at a speed of 62 miles per hour for 15 minutes, how far will it travel?
  ($Distance = rate \times time$)
  a. 9.3 miles
  b. 15.5 miles
  c. 16 miles
  d. 24.8 miles

428. If the speed of light is 3.00×10^8 meters per second, how far would a beam of light travel in 2,000 seconds?
  a. 1.50×10^5 meters
  b. 6.00×10^5 meters
  c. 1.50×10^{11} meters
  d. 6.00×10^{11} meters

429. Which of the following rope lengths is longest?
  (one centimeter = 0.39 inches)
  a. 1 meter
  b. 1 yard
  c. 32 inches
  d. 85 centimeters

430. If Katie's cat weighs 8.5 pounds, what is the approximate weight of the cat in kilograms? (one kilogram = about 2.2 pounds)
  a. 2.9 kilograms
  b. 3.9 kilograms
  c. 8.5 kilograms
  d. 18.7 kilograms

431. If a worker is given a pay increase of $1.25 per hour, what it the total amount of the pay increase for one 40-hour week?
  a. $49.20
  b. $50.00
  c. $50.25
  d. $51.75

432. A teacher purchased a number of supplies to start the new school year. The costs are listed as follows: $12.98, $5.68, $20.64, and $6.76. What is the total cost?
  a. $45.96
  b. $46.06
  c. $46.16
  d. $47.16

▶ **Set 28** (Answers begin on page 182.)

433. A firefighter determines that the length of hose needed to reach a particular building is 175 feet. If the available hoses are 45 feet long, how many sections of hose, when connected together, will it take to reach the building?

  a. 2
  b. 3
  c. 4
  d. 5

434. Approximately how many liters of water will a 10-gallon container hold? (one liter = 1.06 quarts)

  a. 9 liters
  b. 32 liters
  c. 38 liters
  d. 42 liters

435. If one gallon of water weighs 8.35 pounds, a 25-gallon container of water would most nearly weigh

  a. 173 pounds.
  b. 200 pounds.
  c. 209 pounds.
  d. 215 pounds.

436. Roger wants to know if he has enough money to purchase several items. He needs three heads of lettuce, which cost $0.99 each, and two boxes of cereal, which cost $3.49 each. He uses the expression $(3 \times \$0.99) + (2 \times \$3.49)$ to calculate how much the items will cost. Which of the following expressions could also be used?

  a. $3 \times (\$3.49 + \$0.99) - \$3.49$
  b. $3 \times (\$3.49 + \$0.99)$
  c. $(2 + 3) \times (\$3.49 + \$0.99)$
  d. $(2 \times 3) + (\$3.49 \times \$0.99)$

437. If you take recyclables to whichever recycler will pay the most, what is the greatest amount of money you could get for 2,200 pounds of aluminum, 1,400 pounds of cardboard, 3,100 pounds of glass, and 900 pounds of plastic?

Recycler	Aluminum	Cardboard	Glass	Plastic
X	.06/pound	.03/pound	.08/pound	.02/pound
Y	.07/pound	.04/pound	.07/pound	.03/pound

  a. $409
  b. $440
  c. $447
  d. $454

438. If the average person throws away 3.5 pounds of trash every day, how much trash would the average person throw away in one week?

  a. 24.5 pounds
  b. 31.5 pounds
  c. 40.2 pounds
  d. 240 pounds

439. If production line A can produce 12.5 units in an hour, and production line B can produce 15.25 units in an hour, how long will production line A have to work to produce the same amount of units as production line B?

  a. 1 hour
  b. 1.22 hours
  c. 1.50 hours
  d. 1.72 hours

440. Benito earns $12.50 for each hour that he works. If Benito works 8.5 hours per day, five days a week, how much does he earn in a week?
  a. $100.00
  b. $106.25
  c. $406.00
  d. $531.25

441. Des Moines recently received a snow storm that left a total of eight inches of snow. If it snowed at a consistent rate of three inches every two hours, how much snow had fallen in the first five hours of the storm?
  a. 3 inches
  b. 3.3 inches
  c. 5 inches
  d. 7.5 inches

442. A family eats at Joe's Grill and orders the following items from the menu:
  Hamburger $2.95
  Cheeseburger $3.35
  Chicken Sandwich $3.95
  Grilled Cheese $1.95
  If the family orders two hamburgers, one cheeseburger, two chicken sandwiches, and one grilled cheese, what is the total cost of their order?
  a. $15.15
  b. $17.10
  c. $18.05
  d. $19.10

443. If a physical education student burns 8.2 calories per minute while riding a bicycle, how many calories will the same student burn if she rides for 35 minutes?
  a. 246
  b. 286
  c. 287
  d. 387

444. It takes a typing student 0.75 seconds to type one word. At this rate, how many words can the student type in 60 seconds?
  a. 8
  b. 45
  c. 75
  d. 80

445. John's Market sells milk for $2.24 per gallon. Food Supply sells the same milk for $2.08 per gallon. If Mitzi buys two gallons of milk at Food Supply instead of John's, how much will she save?
  a. $0.12
  b. $0.14
  c. $0.32
  d. $0.38

446. An office uses two dozen pencils and $3\frac{1}{2}$ reams of paper each week. If pencils cost five cents each and a ream of paper costs $7.50, how much does it cost to supply the office for a week?
  a. $7.55
  b. $12.20
  c. $26.25
  d. $27.45

447. If a particular woman's resting heartbeat is 72 beats per minute and she is at rest for $6\frac{1}{2}$ hours, about how many times will her heart beat during that period of time?

   a. 4,320

   b. 4,680

   c. 28,080

   d. 43,200

448. Sarah makes 2.5 times more money per hour than Connor does. If Connor earns $7.20 per hour, how much does Sarah make per hour?

   a. $9.70

   b. $14.40

   c. $18.00

   d. $180.00

▶ **Set 29** (Answers begin on page 183.)

449. It takes five-year-old Carlos 1.6 minutes to tie the lace on his right shoe and 1.5 minutes to tie the lace on his left shoe. How many minutes does it take Carlos to tie both shoes?
   a. 2.1 minutes
   b. 3.0 minutes
   c. 3.1 minutes
   d. 4.1 minutes

450. Alicia rode her bicycle a total of 25.8 miles in three days. On average, how many miles did she ride each day?
   a. 8.06 miles
   b. 8.6 miles
   c. 8.75 miles
   d. 8.9 miles

451. If one inch equals 2.54 centimeters, how many inches are there in 254 centimeters?
   a. $\frac{1}{10}$ inch
   b. 10 inches
   c. 100 inches
   d. 1,000 inches

452. Erin ran 6.45 miles on Monday, 5.9 miles on Tuesday, and 6.75 miles on Wednesday. What is the total number of miles Erin ran?
   a. 19.1
   b. 19.05
   c. 17
   d. 13.79

453. Joel's resting heart rate is about 71 beats per minute. If Joel is at rest for 35.2 minutes, about how many times will his heart beat during that period of time?
   a. 2,398.4
   b. 2,408.4
   c. 2,490.3
   d. 2,499.2

454. If one pound of chicken costs $2.79 a pound, how much does 0.89 pound of chicken cost, rounded to the nearest cent?
   a. $2.40
   b. $2.48
   c. $2.68
   d. $4.72

455. On Wednesday morning, Yoder's Appliance Service had a balance of $2,354.82 in its checking account. If the bookkeeper wrote a total of $867.59 worth of checks that day, how much was left in the checking account?
   a. $1,487.23
   b. $1,487.33
   c. $1,496.23
   d. $1,587.33

456. If Nanette cuts a length of ribbon that is 13.5 inches long into four equal pieces, how long will each piece be?
   a. 3.3075 inches
   b. 3.375 inches
   c. 3.385 inches
   d. 3.3805 inches

457. At age six, Zack weighed 40.6 pounds. By age seven, Zack weighed 46.1 pounds. How much weight did he gain in that one year?
a. 4.5 pounds
b. 5.5 pounds
c. 5.7 pounds
d. 6.5 pounds

458. While on a three-day vacation, the Wilsons spent the following amounts on motel rooms: $52.50, $47.99, and $49.32. What is the total amount they spent?
a. $139.81
b. $148.81
c. $148.83
d. $149.81

459. Jake grew 0.6 inch during his senior year in high school. If he was 68.8 inches tall at the beginning of his senior year, how tall was he at the end of the year?
a. 69.0 inches
b. 69.2 inches
c. 69.4 inches
d. 74.8 inches

460. For a science project, Stacy and Tina are measuring the length of two caterpillars. Stacy's caterpillar is 2.345 centimeters long. Tina's caterpillar is 0.0005 centimeter longer. How long is Tina's caterpillar?
a. 2.0345 centimeters
b. 2.3455 centimeters
c. 2.0345 centimeters
d. 2.845 centimeters

461. About how many quarts of water will a 3.25-liter container hold?(one liter = 1.06 quarts)
a. 3.066 quarts
b. 3.045 quarts
c. 3.445 quarts
d. 5.2 quarts

462. Jessica has basic cable television service at a cost of $13.95 per month. If she adds the movie channels, it will cost an additional $5.70 per month. The sports channels cost another $4.89 per month. If Jessica adds the movie channels and the sports channels, what will her total monthly payment be?
a. $23.54
b. $23.55
c. $24.54
d. $34.54

463. The fares collected for one bus on Route G47 on Monday are as follows: Run 1—$419.50, Run 2—$537.00, Run 3—$390.10, Run 4—$425.50. What is the total amount collected?
a. $1,661.10
b. $1,762.20
c. $1,772.10
d. $1,881.00

464. Bart and Sam mow lawns at the same rate. If it takes Bart and Sam about 0.67 hour to mow one half acre lawn together, about how many hours would it take Bart alone to mow five half acre lawns?
a. 3.35 hours
b. 4.35 hours
c. 5.75 hours
d. 6.7 hours

► **Set 30** (Answers begin on page 183.)

465. The town of Crystal Point collected $84,493.26 in taxes last year. This year, the town collected $91,222.30 in taxes. How much more money did the town collect this year?
   a. $6,729.04
   b. $6,729.14
   c. $6,739.14
   d. $7,829.04

466. It took Darren 3.75 hours to drive 232.8 miles. What was his average mile per hour speed?
   a. 62.08 miles per hour
   b. 62.8 miles per hour
   c. 63.459 miles per hour
   d. 71.809 miles per hour

467. Marly has budgeted $100.00 for the week to spend on food. If she buys a beef roast that costs $12.84 and four pounds of shrimp that cost $3.16 per pound, how much of her weekly food budget will she have left?
   a. $74.52
   b. $80.00
   c. $84.00
   d. $86.62

468. Three 15.4-inch pipes are laid end to end. What is the total length of the pipes in feet? (one foot = 12 inches)
   a. 3.02 feet
   b. 3.2 feet
   c. 3.85 feet
   d. 4.62 feet

469. If one ounce equals 28.571 grams, 12.1 ounces is equal to how many grams?
   a. 37.63463 grams
   b. 343.5473 grams
   c. 345.7091 grams
   d. 376.3463 grams

470. Theresa is weighing objects in kilograms. A book weighs 0.923 kilogram; a pencil weighs 0.029 kilogram; an eraser weighs 0.1153 kilogram. What is the total weight of the three objects?
   a. 0.4353 kilogram
   b. 1.0673 kilograms
   c. 1.4283 kilograms
   d. 10.673 kilograms

471. The Cougars played three basketball games last week. Monday's game lasted 113.9 minutes; Wednesday's game lasted 106.7 minutes; and Friday's game lasted 122 minutes. What is the average time, in minutes, for the three games?
   a. 77.6 minutes
   b. 103.2 minutes
   c. 114.2 minutes
   d. 115.6 minutes

472. Ingrid has two pieces of balsa wood. Piece A is 0.724 centimeter thick. Piece B is 0.0076 centimeter thicker than Piece A. How thick is Piece B?
   a. 0.7164 centimeter
   b. 0.7316 centimeter
   c. 0.8 centimeter
   d. 0.08 centimeter

473. Michael has a $20 bill and a $5 bill in his wallet and $1.29 in change in his pocket. If he buys a half gallon of ice cream that costs $4.89, how much money will he have left?
a. $22.48
b. $22.30
c. $21.48
d. $21.40

474. The butcher at Al's Meat Market divided ground beef into eight packages. If each package weighs 0.75 pound and he has 0.04 pound of ground beef left over, how many pounds of ground beef did he start with?
a. 5.064 pounds
b. 5.64 pounds
c. 6.04 pounds
d. 6.4 pounds

475. It is 19.85 miles from Jacqueline's home to her job. If she works five days a week and drives to work, how many miles does Jacqueline drive each week?
a. 99.25 miles
b. 188.5 miles
c. 190.85 miles
d. 198.5 miles

476. Phil and Alice went out to dinner and spent a total of $42.09. If they tipped the waiter $6.25 and the tip was included in their total bill, how much did their meal alone cost?
a. $35.84
b. $36.84
c. $36.74
d. $48.34

477. Antoine earns $8.30 an hour for the first 40 hours he works each week. For every hour he works overtime, he earns 1.5 times his regular hourly wage. If Antoine worked 44 hours last week, how much money did he earn?
a. $365.20
b. $337.50
c. $381.80
d. $547.80

478. The highest temperature in Spring Valley on September 1 was 93.6° F. On September 2, the highest temperature was 0.8° higher than on September 1. On September 3, the temperature was 11.6° lower than on September 2. What was the temperature on September 3?
a. 74° F
b. 82.2° F
c. 82.8° F
d. 90° F

479. A survey has shown that a family of four can save about $40 a week if they purchase generic items rather than brand-name ones. How much can a particular family save over six months? (one month = 4.3 weeks)
a. $1,032
b. $1,320
c. $1,310
d. $1,300

480. The Benton High School girls' relay team ran the mile in 6.32 minutes in April. By May, they were able to run the same race in 6.099 minutes. By how many minutes had their time improved?

  a. 0.221 minute
  b. 0.339 minute
  c. 0.467 minute
  d. 0.67 minute

4 ▶ Percentages

The following 10 sets of problems deal with percentages, which, like decimals, are a special kind of fraction. Percentages have many everyday uses, from figuring the tip in a restaurant to understanding complicated interest and inflation rates. This section will give you practice in working with the relationship between percents, decimals, and fractions, and with changing one into another. You will also practice ratios and proportions, which are similar to percentages.

▶ **Set 31** (Answers begin on page 185.)

481. 6% =
   a. 6.0
   b. 0.6
   c. 0.06
   d. 0.006

482. 9% =
   a. 0.9
   b. 9.0
   c. 0.09
   d. 0.009

483. 37% =
   a. 0.037
   b. 0.37
   c. 3.7
   d. 3.07

484. 3.04% =
   a. 3.04
   b. 0.304
   c. 0.0304
   d. 304

485. 500% =
   a. 0.05
   b. 0.5
   c. 5.0
   d. 50.0

486. 0.06 =
   a. 0.60%
   b. 6.0%
   c. 60.0%
   d. 600%

487. $4\frac{1}{5}$% =
   a. 420%
   b. 0.420%
   c. 4.20%
   d. 42%

488. 20% =
   a. 0.0002
   b. 0.002
   c. 0.02
   d. 0.2

489. $\frac{1}{5}$% =
   a. 0.002%
   b. 0.02%
   c. 0.20%
   d. 20.0%

490. $\frac{1}{5}$ =
   a. 20%
   b. 0.20%
   c. 0.020%
   d. 0.002%

491. 0.66 is equal to
   a. 66%
   b. 6.6%
   c. 0.66%
   d. 0.066%

492. 25% converted to a fraction =
   a. $\frac{25}{100}$
   b. $\frac{1}{25}$
   c. $\frac{25}{1}$
   d. $\frac{25}{25}$

493. 625% converted to a mixed number =
a. $62\frac{1}{4}$
b. $6\frac{1}{4}$
c. $0.6\frac{1}{4}$
d. 0.625

494. 32% converted to a fraction =
a. $\frac{1}{32}$
b. $\frac{8}{32}$
c. $\frac{8}{25}$
d. $\frac{1}{25}$

495. 70% of 600 =
a. 420
b. 180
c. 480
d. 320

496. 40% of 240 =
a. 144
b. 96
c. 120
d. 98

▶ **Set 32** (Answers begin on page 185.)

497. 53% of 765 =
- **a.** 395.55
- **b.** 405.54
- **c.** 359.55
- **d.** 405.45

498. 200% of 40 =
- **a.** 4
- **b.** 60
- **c.** 80
- **d.** 100

499. 30% of 39 =
- **a.** 27.3
- **b.** 11.7
- **c.** 2.73
- **d.** 1.17

500. 72% converted to a decimal =
- **a.** 0.0072
- **b.** 0.072
- **c.** 0.72
- **d.** 7.2

501. Which of the following is 16% of 789?
- **a.** 126.24
- **b.** 12.624
- **c.** 662.76
- **d.** 66.276

502. What percentage of 12,000 is 216?
- **a.** 1,800%
- **b.** 180%
- **c.** 18%
- **d.** 1.8%

503. What percentage of 700 is 1,225?
- **a.** 57%
- **b.** 60%
- **c.** 125%
- **d.** 175%

504. What is $4\frac{1}{4}$% of 574, rounded to the nearest tenth?
- **a.** 24.4
- **b.** 24.3
- **c.** 20
- **d.** 30

505. 62.5% is equal to
- **a.** $\frac{1}{16}$
- **b.** $\frac{5}{8}$
- **c.** $6\frac{1}{4}$
- **d.** $6\frac{2}{5}$

506. Convert $\frac{6}{80}$ to a percentage.
- **a.** 0.75%
- **b.** 0.075%
- **c.** 7.5%
- **d.** 75%

507. Approximately how much money is a 20% tip on a restaurant bill of $16?
- **a.** $0.32
- **b.** $3.20
- **c.** $6.40
- **d.** $12.80

508. Approximately how much money is a 15% tip on a restaurant bill of $24?
- **a.** $2.40
- **b.** $3.20
- **c.** $3.60
- **d.** $4.80

509. What is 53.2% of 18?
    a. 0.8424
    b. 0.9579
    c. 8.424
    d. 9.576

510. What is 0.4% of 30?
    a. 12
    b. 0.12
    c. 1.2
    d. 0.012

511. What is 42% of 6?
    a. 0.00252
    b. 0.025
    c. 2.52
    d. 0.252

512. 60% is equivalent to
    a. 0.6 and $\frac{6}{10}$
    b. 6.0 and $\frac{6}{10}$
    c. 0.6 and $\frac{10}{6}$
    d. 0.06 and $\frac{6}{6}$

▶ **Set 33** (Answers begin on page 186.)

513. 25% of what number is 18?

    a. 8

    b. 36

    c. 54

    d. 72

514. Which of the following phrases means "percent"?

    a. *per part*

    b. *per 100 parts*

    c. *per fraction*

    d. *per decimal*

515. Which of the following terms is best described as "a comparison of two numbers"?

    a. variable

    b. coefficient

    c. ratio

    d. radical

516. Which of the following is equal to 0.34?

    a. 34

    b. $\frac{34}{100}$

    c. $\frac{34}{34}$

    d. $\frac{100}{34}$

517. Which operation does the fraction bar in a fraction represent?

    a. addition

    b. subtraction

    c. multiplication

    d. division

518. What percent of 50 is 12?

    a. 4%

    b. 14%

    c. 24%

    d. 0.4%

519. 21.12 is 12% of what number?

    a. 176

    b. 2.5344

    c. 253.44

    d. 12

520. Change $\frac{5}{20}$ to a percent.

    a. 5%

    b. 25%

    c. 20%

    d. 100%

521. 0.8 =

    a. 8%

    b. 0.8%

    c. 80%

    d. 800%

522. $\frac{1}{4} \times 100\% =$

    a. 25%

    b. 0.25%

    c. 2%

    d. 2.5%

523. Membership dues at Arnold's Gym are $53 per month this year, but were $50 per month last year. What was the percentage increase in the gym's prices?

    a. 5.5%

    b. 6.0%

    c. 6.5%

    d. 7.0%

524. Lucille spent 12% of her weekly earnings on DVDs and deposited the rest into her savings account. If she spent $42 on DVDs, how much did she deposit into her savings account?
a. $42
b. $308
c. $318
d. $350

525. Yetta just got a raise of $3\frac{1}{4}$%. Her original salary was $30,600. How much does she make now?
a. $30,594.50
b. $31,594.50
c. $32,094.50
d. $32,940.50

Use the following passage to answer question 526.

Basic cable television service, which includes 16 channels, costs $15 a month. The initial labor fee to install the service is $25. A $65 deposit is required but will be refunded within two years if the customer's bills are paid in full. Other cable services may be added to the basic service: The movie channels are $9.40 a month; the news channels are $7.50 a month; the arts channels are $5.00 a month; and the sports channels are $4.80 a month.

526. A customer's cable television bill totaled $20 per month. Using the previous passage, what portion of the bill was for basic cable service?
a. 25%
b. 33%
c. 50%
d. 75%

527. Gloria has finished reading 30% of a 340-page novel. How many pages has she read?
a. 102
b. 103
c. 105
d. 113

528. Ten students from the 250-student senior class at Jefferson High School have received full college scholarships. What percentage of the senior class received full college scholarships?
a. 2%
b. 4%
c. 10%
d. 25%

▶ **Set 34** (Answers begin on page 187.)

529. Kirsten's dinner at a local restaurant cost
$13.85. If she wants to leave the server a tip
that equals 20% of the cost of her dinner, how
much of a tip should she leave?
   a. $2.00
   b. $2.67
   c. $2.77
   d. $3.65

530. This month, attendance at the baseball stadium
was 150% of the attendance last month. If
attendance this month was 280,000, what was
the attendance last month, rounded to the
nearest whole number?
   a. 140,000
   b. 176,670
   c. 186,667
   d. 205,556

531. Edward purchased a house for $70,000. Five
years later, he sold it for an 18% profit. What
was his selling price?
   a. $82,600
   b. $83,600
   c. $85,500
   d. $88,000

532. The price of gasoline drops from $1.00 per gal-
lon to $0.95 per gallon. What is the percent of
decrease?
   a. 2%
   b. 3%
   c. 4%
   d. 5%

533. A certain power company gives a $1\frac{1}{2}$% discount
if a customer pays the bill at least 10 days
before the due date. If Inez pays her $48.50 bill
10 days early, how much money will she save,
rounded to the nearest cent?
   a. $0.49
   b. $0.73
   c. $1.50
   d. $7.28

534. Each year, on average, 24 of the 480 students at
a certain high school are members of the science team. What percentage of students is on
the science team?
   a. 0.05%
   b. 0.5%
   c. 5%
   d. 15%

535. Marty and Phyllis arrive late for a movie and
miss 10% of it. The movie is 90 minutes long.
How many minutes did they miss?
   a. 10 minutes
   b. 9 minutes
   c. 8 minutes
   d. 7 minutes

536. Jeremy ate three ounces of a 16-ounce carton of
ice cream. What percentage of the carton did
he eat?
   a. 18.75%
   b. 17.25%
   c. 19.50%
   d. 16.75%

537. Wendy's pay is $423.00; 19% of that is subtracted for taxes. How much is her take-home pay?
a. $404.44
b. $355.46
c. $342.63
d. $455.45

538. In the accounting department at a university, 60% of the students are women and 250 of the students are men. How many accounting students are there in all?
a. 40
b. 375
c. 675
d. 625

539. A ticket to an evening movie at a theater costs $7.50. The cost of popcorn at the concession stand is equal to 80% of the cost of a ticket. How much does the popcorn cost?
a. $5.50
b. $6.00
c. $6.50
d. $7.00

540. This week the stock market closed at 8,990 points. Last week it closed at 7,865 points. What was the percentage of increase this week, rounded to the nearest whole number percent?
a. 14%
b. 15%
c. 85%
d. 86%

541. Veronique borrowed $10,000 from her uncle Kevin and agreed to pay him 4.5% interest, compounded yearly. How much interest did she owe the first year?
a. $145
b. $100
c. $400
d. $450

542. Rookie police officers have to buy duty shoes at the full price of $84.50, but officers who have served at least a year get a 15% discount. Officers who have served at least three years get an additional 10% off the discounted price. How much does an officer who has served at least three years have to pay for shoes?
a. $63.78
b. $64.65
c. $71.83
d. $72.05

543. At the city park, 32% of the trees are oaks. If there are 400 trees in the park, how many trees are NOT oaks?
a. 128
b. 272
c. 278
d. 312

544. The town of Centerville spends 15% of its annual budget on its public library. If Centerville spent $3,000 on its public library this year, what was its annual budget this year?
a. $15,000
b. $20,000
c. $35,000
d. $45,000

▶ Set 35 (Answers begin on page 188.)

545. Of the 1,200 videos available for rent at a certain video store, 420 are comedies. What percent of the videos are comedies?
  a. $28\frac{1}{2}\%$
  b. 30%
  c. 32%
  d. 35%

546. Toby, a golden retriever, gained 5.1 pounds this month. If Toby now weighs 65.1 pounds, what is the percent increase in Toby's weight?
  a. 5.9%
  b. 6%
  c. 8.5%
  d. 9.1%

547. Glenda bought a sofa-sleeper at a 10%-off sale and paid the sale price of $575.00. What was the price, rounded to the nearest cent, of the sofa-sleeper before the sale?
  a. $585.00
  b. $587.56
  c. $633.89
  d. $638.89

548. Nathan saves $5\frac{1}{4}\%$ of his weekly salary. If Nathan earns $380 per week, how much does he save each week?
  a. $19.95
  b. $20.52
  c. $21.95
  d. $25.20

549. Erin has completed 70% of her homework. If she has been doing homework for 42 minutes, how many more minutes does she have left to work?
  a. 15 minutes
  b. 18 minutes
  c. 20.5 minutes
  d. 28 minutes

550. Terrill cuts a piece of rope into three pieces. One piece is eight feet long, one piece is seven feet long, and one piece is five feet long. The shortest piece of rope is what percent of the original length before the rope was cut?
  a. 4%
  b. 18.5%
  c. 20%
  d. 25%

551. Pam's monthly food budget is equal to 40% of her monthly house payment. If her food budget is $200 a month, how much is her house payment each month?
  a. $340
  b. $400
  c. $500
  d. $540

552. If container A holds eight gallons of water, and container B holds 12% more than container A, how many gallons of water does container B hold?
  a. 8.12 gallons
  b. 8.48 gallons
  c. 8.96 gallons
  d. 9.00 gallons

553. Thirty-five cents is what percent of $1.40?

    a. 25%

    b. 40%

    c. 45%

    d. 105%

554. The White-Bright Toothbrush Company hired 30 new employees. This hiring increased the company's total workforce by 5%. How many employees now work at White-Bright?

    a. 530

    b. 600

    c. 605

    d. 630

555. Matthew had 200 baseball cards. He sold 5% of the cards on Saturday and 10% of the remaining cards on Sunday. How many cards are left?

    a. 170

    b. 171

    c. 175

    d. 185

556. Kate spent 45% of the money that was in her wallet. If she spent $9.50, how much money was in her wallet to begin with? Round to the nearest cent.

    a. $4.28

    b. $4.38

    c. $14.49

    d. $21.11

557. Steve earned a $4\frac{3}{4}$% pay raise. If his salary was $27,400 before the raise, how much was his salary after the raise?

    a. $27,530.15

    b. $28,601.50

    c. $28,701.50

    d. $29,610.50

558. The temperature in Sun Village reached 100° or more about 15% percent of the past year. About how many days did the temperature in Sun Village climb to 100° or higher? (one year = 365 days) Round your answer.

    a. 45

    b. 54

    c. 55

    d. 67

559. A certain radio station plays classical music during 20% of its airtime. If the station is on the air 24 hours a day, how many hours each day is the station NOT playing classical music?

    a. 8.0 hours

    b. 15.6 hours

    c. 18.2 hours

    d. 19.2 hours

560. In order to pass a certain exam, candidates must answer 70% of the test questions correctly. If there are 70 questions on the exam, how many questions must be answered correctly in order to pass?

    a. 49

    b. 52

    c. 56

    d. 60

▶ **Set 36** (Answers begin on page 189.)

561. Of 150 people polled, 105 said they rode the city bus at least three times per week. How many people out of 100,000 could be expected to ride the city bus at least three times each week?
   a. 55,000
   b. 70,000
   c. 72,500
   d. 75,000

562. Of 360 students polled, 150 participate in extracurricular activities. Approximately what percent of the students do NOT participate in extracurricular activities?
   a. 32%
   b. 42%
   c. 52%
   d. 58%

563. Kate earns $26,000 a year. If she receives a 4.5% salary increase, how much will she earn?
   a. $26,450
   b. $27,170
   c. $27,260
   d. $29,200

564. A sprinkler system installed in a home that is under construction will cost about 1.5% of the total building cost. The same system, installed after the home is built, is about 4% of the total building cost. How much would a homeowner save by installing a sprinkler system in a $150,000 home while the home is still under construction?
   a. $600
   b. $2,250
   c. $3,750
   d. $6,000

565. Out of 100 shoppers polled, 80 said they buy fresh fruit every week. How many shoppers out of 30,000 could be expected to buy fresh fruit every week?
   a. 2,400
   b. 6,000
   c. 22,000
   d. 24,000

566. The novel that Lynn is reading contains 435 pages. So far, she has read 157 pages. Approximately what percent of the novel has she read?
   a. 34%
   b. 36%
   c. 44%
   d. 64%

567. While planning for an event, 40% of the members of a committee attended a meeting. If there are a total of 75 members on the committee, how many of the members attended this meeting?
   a. 25
   b. 30
   c. 35
   d. 40

568. A pump installed on a well can pump at a maximum rate of 100 gallons per minute. If the town requires 15,000 gallons of water in one day, how long would the pump have to continuously run at 75% of its maximum rate to meet the town's need?
   a. 112.5 minutes
   b. 150 minutes
   c. 200 minutes
   d. 300 minutes

569. A company makes several items, including filing cabinets. One-third of their business consists of filing cabinets, and 60% of their filing cabinets are sold to businesses. What percent of their total business consists of filing cabinets sold to businesses?

a. 20%

b. 33%

c. 40%

d. 60%

570. A gram of fat contains nine calories. An 1,800-calorie diet allows no more than 20% calories from fat. How many grams of fat are allowed in that diet?

a. 40 grams

b. 90 grams

c. 200 grams

d. 360 grams

571. Mr. Beard's temperature is 98° Fahrenheit. What is his temperature in degrees Celsius rounded to the nearest tenth? $C = \frac{5}{9}(F - 32)$

a. 35.8° C

b. 36.7° C

c. 37.6° C

d. 31.1° C

572. After having a meal at a restaurant, Jared is charged 9% of the cost of the meal in sales tax. In addition, he wants to leave 15% of the cost of the meal before tax as a tip for the server. If the meal costs $24.95 before tax and the tip, what is the total amount he needs to pay for the meal, tax, and tip?

a. $5.99

b. $28.69

c. $30.94

d. $47.41

573. Of the 4,528 students at a college, 1,132 take a language class. What percentage of the students takes a language class?

a. 4%

b. 20%

c. 25%

d. 40%

Use the following pie chart to answer questions 574 and 575.

Harold's Monthly Budget

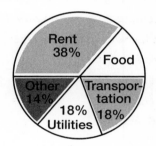

574. What should be the percent label for food?

a. 12%

b. 18%

c. 28%

d. 38%

575. If Harold's monthly income is $2,450, how much does he spend on rent each month?

a. $686

b. $735

c. $882

d. $931

576. A shirt that regularly costs $34 is marked down 15%. What is the sale price of the shirt?

a. $19.00

b. $28.90

c. $29.50

d. $33.85

▶ **Set 37** (Answers begin on page 190.)

577. During exercise, a person's heart rate should be between 60% and 90% of the difference between 220 and the person's age. According to this guideline, what should a 30-year-old person's maximum heart rate be during exercise?
  a. 114 beats per minute
  b. 132 beats per minute
  c. 171 beats per minute
  d. 198 beats per minute

578. A class enrollment of 30 students was increased by 20%. What is the new class size?
  a. 36 students
  b. 39 students
  c. 42 students
  d. 50 students

579. A dishwasher on sale for $279 has an original price of $350. What is the percent of discount, rounded to the nearest percent?
  a. 20%
  b. 25%
  c. 26%
  d. 71%

580. For half of all allergy sufferers, a prescription reduces the number of symptoms by 50%. What percentage of all allergy symptoms can be eliminated by this prescription?
  a. 25%
  b. 50%
  c. 75%
  d. 100%

581. There are 171 men and 129 women enrolled in a program. What percentage of this program is made up of women?
  a. 13%
  b. 43%
  c. 57%
  d. 75%

Use the following pie chart to answer question 582.

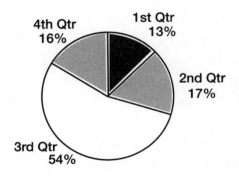

Sales for 2009

582. The chart shows quarterly sales for Cool-Air's air-conditioning units. Which of the following combinations contributed 70% to the total?
  a. first and second quarters
  b. second and third quarters
  c. second and fourth quarters
  d. third and fourth quarters

583. In the Valleyview School District last year, 315 students played on a sports team. Of those, 40% were also involved in the music program. How many students on sports teams were NOT involved in the music program?
  a. 60
  b. 126
  c. 189
  d. 216

584. A floor plan is drawn to scale so that one quarter inch represents two feet. If a hall on the plan is four inches long, how long will the actual hall be when it is built?
  a. 2 feet
  b. 8 feet
  c. 16 feet
  d. 32 feet

585. The markup on a pair of sneakers is 150%. If the sneakers originally cost $45, what is the price after the markup?
  a. $22.50
  b. $57.50
  c. $67.50
  d. $112.50

586. An insurance policy pays 80% of the first $20,000 of a certain patient's medical expenses, 60% of the next $40,000, and 40% of the $40,000 after that. If the patient's total medical bill is $92,000, how much will the policy pay?
  a. $36,800
  b. $49,600
  c. $52,800
  d. $73,600

587. A pump installed on a well can pump at a maximum rate of 100 gallons per minute. If the pump runs at 50% of its maximum rate for six hours a day, how much water is pumped in one day?
  a. 3,000 gallons
  b. 18,000 gallons
  c. 36,000 gallons
  d. 72,000 gallons

588. Melissa can grade five of her students' papers in one hour. Joe can grade four of the same papers in one hour. If Melissa works for three hours grading, and Joe works for two hours, what percentage of the 50 students' papers will be graded?
  a. 44%
  b. 46%
  c. 52%
  d. 54%

589. Dimitri has 40 math problems to do for homework. If he does 40% of the assignment in one hour, how long will it take for Dimitri to complete the whole assignment?
  a. 1.5 hours
  b. 2.0 hours
  c. 2.5 hours
  d. 3.0 hours

590. The population of Smithtown increases at a rate of 3% annually. If the population is currently 2,500, what will the population be at the same time next year?
  a. 2,530
  b. 2,560
  c. 2,575
  d. 2,800

591. The Chen family traveled 75 miles to visit relatives. If they traveled 57.8% of the way before they stopped at a gas station, how far was the gas station from their relatives' house? Round your answer to the nearest $\frac{1}{3}$ mile.

a. $31\frac{2}{3}$ miles

b. $32\frac{2}{3}$ miles

c. 35 miles

d. $38\frac{1}{3}$ miles

592. A recent survey polled 2,500 people about their reading habits. The results are as follows:

Reading Survey	
Books per month	**Percentage**
0	13
1–3	27
4–6	32
>6	28

How many people surveyed had read books in the last month?

a. 700

b. 1,800

c. 1,825

d. 2,175

▶ **Set 38** (Answers begin on page 191.)

593. A machine on a production line produces parts that are not acceptable by company standards 4% of the time. If the machine produces 500 parts, how many will be defective?
   a. 8
   b. 10
   c. 16
   d. 20

594. Thirty percent of the high school is involved in athletics. If 15% of the athletes play football, what percentage of the whole school plays football?
   a. 4.5%
   b. 9.0%
   c. 15%
   d. 30%

595. Twenty percent of the people at a restaurant selected the dinner special. If 40 people did not select the special, how many people are eating at the restaurant?
   a. 10
   b. 20
   c. 40
   d. 50

596. An average of 90% is needed on five tests to receive an A in a class. If a student received scores of 95, 85, 88, and 84 on the first four tests, what will the student need on the fifth test to get an A?
   a. 92
   b. 94
   c. 96
   d. 98

597. Only $\frac{3}{8}$ of the students in Charlie's class bring their lunch to school each day. What percentage of the students brings their lunch?
   a. 12.5%
   b. 24.5%
   c. 34.5%
   d. 37.5%

598. What is 90% of 90?
   a. 9
   b. 18
   c. 81
   d. 89

599. The City Bus Department operates 200 bus routes. Of these, $5\frac{1}{2}$% are express routes. How many express routes are there?
   a. 11
   b. 15
   c. 22
   d. 25

600. Twelve is 20% of what number?

   a. 5

   b. 20

   c. 60

   d. 240

601. A certain baseball player gets a hit about three out of every 12 times he is at bat. What percentage of the times he is at bat does he get a hit?

   a. 25%

   b. 32%

   c. 35%

   d. 40%

602. Herschel has worked 40% of his eight-hour shift at the widget factory. How many hours has he worked?

   a. 3 hours

   b. 3.2 hours

   c. 3.4 hours

   d. 3.5 hours

603. Eighteen of the 45 guests at a banquet ordered seafood for dinner. What percent of the guests ordered a seafood meal?

   a. 18%

   b. 33%

   c. 40%

   d. 45%

604. Paolo estimates that his college expenses this year will be $26,000; however, he wishes to add 15% to that amount in case of emergency. How much should Paolo try to add to his college fund?

   a. $1,733

   b. $2,900

   c. $3,000

   d. $3,900

605. Because it was her birthday, Patty spent 325% more than she usually spends for lunch. If she usually spends $4.75 per day for lunch, how much did she spend today?

   a. $9.50

   b. $15.44

   c. $20.19

   d. $32.50

606. Eddie weighed 230 pounds in 2008. Today he weighs 196 pounds. About what percentage of weight has he lost since 2008?

   a. 15%

   b. 24%

   c. 30%

   d. 34%

607. A discount retail store gives a 5% discount to senior citizens. If a customer who is a senior citizen wants to purchase an item that costs $45, what is the final cost after the discount is applied?

a. $2.25

b. $42.75

c. $47.25

d. $50.00

608. Pahana took a 250-question physics test and got 94% of the answers correct. How many questions did he answer correctly?

a. 240

b. 235

c. 230

d. 225

▶ **Set 39** (Answers begin on page 192.)

609. What percent of the figure is shaded?

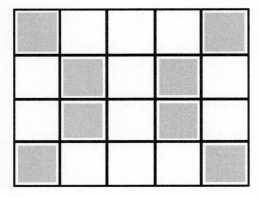

    a. 20%

    b. 25%

    c. 40%

    d. 50%

610. After taxes are taken out, Tim's net paycheck is $363.96. If 28% of his pay goes to taxes, what is his gross pay before taxes?

    a. $101.91

    b. $1,299.86

    c. $465.87

    d. $505.50

611. At a clothing store, 8% of the dresses are designer dresses and the rest are not. If there are 300 dresses at the boutique, how many are NOT designer dresses?

    a. 136

    b. 276

    c. 296

    d. 292

612. Ben spends 15% of his weekly budget on transportation. This week he spent $48 on transportation. What was his weekly budget this week?

    a. $720

    b. $520

    c. $305

    d. $320

613. There are 26 pies in the county fair pie contest this year. Of these, four are peach. About what percent of the pies are peach?

    a. 4%

    b. 9%

    c. 12%

    d. 15%

614. Michelle earned $4 last week selling baked goods. If she earns $36 this week, what is the percent of increase over last week's total?

    a. 32%

    b. 36%

    c. 80%

    d. 800%

615. A sweater that originally cost $34.99 is marked up 25%. How much does the sweater cost after the markup?

    a. $8.25

    b. $8.75

    c. $2.50

    d. $43.74

616. Martha bought a lawnmower on sale for 35% off and paid $298.35. What was the price of the lawnmower before it was marked down?
a. $193.93
b. $333.35
c. $350.50
d. $459.00

617. Barbara buys six dolls and saves $3\frac{1}{2}$% of the total price by buying in bulk. If each doll originally costs $300, how much does Barbara save in total?
a. $10.50
b. $54.00
c. $63.00
d. $75.00

618. Martin, a motel housekeeper, has finished cleaning about 40% of the 32 rooms he's been assigned. About how many more rooms does he have left to clean?
a. 29
b. 25
c. 21
d. 19

619. Garth cuts a piece of rope into three pieces. One piece is five inches long, one piece is four inches long, and one piece is three inches long. The longest piece of rope is approximately what percent of the original length before the rope was cut?
a. 33%
b. 42%
c. 50%
d. 55%

620. Robert's monthly utility bill is equal to 60% of his monthly rent, which is $500 per month. How much is Robert's utility bill each month?
a. $560
b. $440
c. $300
d. $200

621. Forty cents is what percent of $1.30?
a. 40%
b. 31%
c. 20%
d. 11%

622. In the high school choir, the ratio of males to females is three to two. What percentage of the choir is female?
a. 40%
b. 60%
c. 67%
d. 150%

623. In a city last year, the ratio of rainy days to sunny days was three to four. Approximately what percent of days were sunny last year?
a. 43%
b. 57%
c. 75%
d. 133%

624. Toby had a bag of 48 candies. He ate 25% of the candies for dessert one evening and then shared 25% of the remaining candies with his friends. About how many candies are now left?
a. 9
b. 12
c. 24
d. 27

▶ **Set 40** (Answers begin on page 193.)

625. Hilga and Jerome leave from different points walking directly toward each other. Hilga walks $2\frac{1}{2}$ miles per hour, and Jerome walks 4 miles per hour. If they meet in $2\frac{1}{2}$ hours, how far apart were they?

a. 9 miles

b. 13 miles

c. $16\frac{1}{4}$ miles

d. $18\frac{1}{2}$ miles

626. Fifteen percent of 40 beagles offered dry or canned dog food chose dry dog food. How many beagles chose canned food?

a. 34

b. 32

c. 8

d. 6

627. Lucy and Ethel dropped one of every eight chocolates from the conveyer belt onto the floor. What percentage of chocolates did they drop on the floor?

a. 12.5%

b. 13.2%

c. 14.5%

d. 15.2%

628. Philbert gave the pizza delivery person a tip of $4, which was 20% of his total bill for the pizza he ordered. How much did the pizza cost?

a. $30

b. $25

c. $15

d. $20

629. A popular news show broadcasts for an hour. During that time, there are 15 minutes of commercials. What percentage of the hour is devoted to commercials?

a. 15%

b. 35%

c. 25%

d. 45%

630. Manuelita had 75 stuffed animals. Her grandmother gave 15 of them to her. What percentage of the stuffed animals did her grandmother give her?

a. 20%

b. 15%

c. 25%

d. 10%

631. At the garage, Xena's bill to have her car repaired was $320. The total charge for labor was $80. What percentage of the bill was for labor?

a. 15%

b. 20%

c. 25%

d. 30%

632. Roger has completed 78% of his 200-page thesis. How many pages has he written?

a. 150

b. 156

c. 165

d. 160

633. A house is valued at $185,000, in a community that assesses property at 85% of value. If the tax rate is $24.85 per thousand dollars assessed, how much is the property tax bill?

a. $1,480
b. $1,850
c. $3,907.66
d. $4,597.25

634. The Owens family is buying a $100,000 house with 5% down payment at closing and a mortgage for $95,000. They must pay an upfront mortgage insurance premium (MIP) at closing of 2.5% of the mortgage amount. In addition, they must pay miscellaneous other closing costs of $1,000. How much money will they need at the closing?

a. $29,750
b. $8,500
c. $8,375
d. $6,000

635. After paying a commission to his broker of 7% of the sale price, a seller receives $103,000 for his house. How much did the house sell for?

a. $95,790
b. $110,000
c. $110,420
d. $110,753

636. Simon Hersch, a salesperson associated with broker Bob King, lists a house for $115,000, with 6% commission due at closing, and finds a buyer. Bob King's practice is that 45% of commissions go to his office and the rest to the salesperson. How much will Simon Hersch make on the sale?

a. $3,105
b. $3,240
c. $3,795
d. $3,960

637. A storekeeper leases her store building for the following amount: $1,000 per month rent, $\frac{1}{12}$ of the $18,000 annual tax bill, and 3% of the gross receipts from her store. If the storekeeper takes in $75,000 in one month, what will her lease payment be?

a. $4,750
b. $3,750
c. $3,250
d. $7,400

638. The appraised value of a property is $325,000, assessed at 90% of its appraisal. If the tax rate for the year is $2.90 per thousand of assessment, how much are the taxes for the first half of the year?

a. $471.25
b. $424.13
c. $942.50
d. $848.25

639. The seller accepts an offer to purchase a house for $395,000. After paying a brokerage fee that is 5.5% of the sale price, other settlement fees totaling 4% of the sale price, and paying off a loan of $300,000, what are the seller's net proceeds?

    a. $57,475

    b. $73,275

    c. $58,344

    d. $373,275

640. In constructing her financial report, Noel Carpenter estimates that her real estate holdings have appreciated by 18% since purchase. If the original value was $585,000, what would her balance sheet show now?

    a. $690,300

    b. $585,000

    c. $479,700

    d. $526,500

5 ▶ Algebra

Basic algebra problems, such as those in the following 11 sets, ask you to solve equations in which one element, or more than one, is unknown and generally indicated by a letter of the alphabet (often either *x* or *y*). In doing the following problems, you will get practice in isolating numbers on one side of the equation and unknowns on the other, thus finding the replacement for the unknown. You'll also practice expressing a problem in algebraic form, particularly when you get to the word problems. Other skills you will practice include working with exponents and roots, factoring, and dealing with polynomial expressions.

▶ **Set 41** (Answers begin on page 195.)

641. If $6x = 42$, then x is
   a. 7
   b. 8
   c. 48
   d. 36

642. If $[\frac{x}{72}] = 2$, then x is
   a. 9
   b. 36
   c. 144
   d. 72

643. If $5a + 50 = 150$, then a is
   a. 10
   b. 20
   c. 30
   d. 40

644. The product of a number and its square is 729. What is the number?
   a. 9
   b. 364.5
   c. 18
   d. 182.25

645. Sixteen less than six times a number is 20. What is the number?
   a. 12
   b. 10
   c. 8
   d. 6

646. A certain number when added to 25% of itself is 125. What is the number?
   a. 25
   b. 50
   c. 75
   d. 100

647. Eight more than $\frac{1}{7}$ of 42 is
   a. 12
   b. 13
   c. 14
   d. 15

648. The square root of a number is three times 2^2. What is the number?
   a. 100
   b. 121
   c. 144
   d. 169

649. What is the sum of the first four prime numbers?
   a. 17
   b. 18
   c. 19
   d. 20

650. If $9b + 6b = 15$, then b is
   a. 8
   b. 4
   c. 2
   d. 1

651. What is the smallest prime number?

    a. 0

    b. 1

    c. 2

    d. 3

652. When both six and nine are added to a number, the sum is 49. What is the number?

    a. 15

    b. 34

    c. 43

    d. 21

653. Forty-two is about what percent of 80?

    a. 0.51%

    b. 0.32%

    c. 32%

    d. 51%

654. The product of nine and one-third of a number is 81. Find the number.

    a. 162

    b. 27

    c. 3

    d. 90

655. Twice the product of the first two multiples of this number is 196. Find the number.

    a. 19

    b. 29

    c. 7

    d. 8

656. Five more than 20% of a number is 5^2. Find the number.

    a. 50

    b. 60

    c. 70

    d. 100

▶ **Set 42** (Answers begin on page 195.)

657. Seventy-five together with $\frac{12}{6}$ of a number is 135. What is the number?

a. 20

b. 30

c. 40

d. 50

658. The statement $x + y = y + x$ is an example of what number property?

a. Commutative Property of Addition

b. Associative Property of Addition

c. Identity Property of Addition

d. Inverse Property of Addition

659. What is the value of the expression $xy - 6z$, when $x = -3$, $y = 6$, and $z = -5$?

a. −48

b. 48

c. −12

d. 12

660. Thirty-four more than a certain number is 98. What is the number?

a. 84

b. 132

c. 64

d. 17

661. A number is three times larger when 10 is added to it. What is the number?

a. 33

b. 7

c. 13

d. 5

662. Which value of a will make this number sentence false? $a \leq 5$

a. 0

b. −3

c. 5

d. 6

663. In the equation $4p - 10 - 2p = 16$, what is p equal to?

a. 2

b. 6

c. 13

d. 26

664. Fifty plus three times a number is 74. What is the number?

a. 2

b. 4

c. 6

d. 8

665. The product of two and four more than three times a number is 20. What is the number?

a. 2

b. 16

c. 44

d. 87

666. To solve for an unknown in an equation, you must always

a. add it in.

b. subtract it from.

c. isolate it on one side.

d. eliminate the inequality.

667. What is the value of x when $y = 8$ and $x = 4 + 6y$?

a. 48
b. 52
c. 24
d. 36

668. What is the value of the expression $\frac{xy + yz}{xy}$ when $x = 1$, $y = 3$, and $z = 6$?

a. 3
b. 7
c. 12
d. 21

669. Which of the following is an example of the Associative Property of Multiplication?

a. $a(b + c) = ab + ac$
b. $ab = ba$
c. $a(bc) = (ab)c$
d. $a \times 1 = a$

670. What is the greatest common factor of the following monomials: $3x^2$, $12x$, $6x^3$

a. 12
b. $3x$
c. $6x$
d. $3x^2$

671. Which value of x will make this number sentence true: $x + 25 \leq 13$

a. -13
b. -11
c. 12
d. 38

672. $\frac{x}{4} + \frac{3x}{4} =$

a. $\frac{1}{2}x$
b. $\frac{x^3}{4}$
c. 1
d. x

▶ **Set 43** (Answers begin on page 196.)

673. Which of the following lists three consecutive even integers whose sum is 30?

    a. 9, 10, 11

    b. 8, 10, 12

    c. 9, 11, 13

    d. 10, 12, 14

674. Which of the following represents the factors of the trinomial $x^2 + 6x + 9$?

    a. $x(x + 9)$

    b. $(x + 6)(x + 9)$

    c. $(x + 3)(x + 6)$

    d. $(x + 3)^2$

675. If $\frac{2x}{16} = \frac{12}{48}$, what is x?

    a. 2

    b. 3

    c. 4

    d. 5

676. Which value of x will make the following inequality true: $12x - 1 < 35$

    a. 2

    b. 3

    c. 4

    d. 5

677. Which of the following is a simplification of $(x^2 + 4x + 4) \div (x + 2)$?

    a. $x - 2$

    b. $x + 4$

    c. $x^2 + 3x + 2$

    d. $x + 2$

678. Which of the following is equivalent to $x^3 + 6x$?

    a. $x(x^2 + 6)$

    b. $x(x + 6)$

    c. $x(x^2 + 6x)$

    d. $x^2(x + 6)$

679. When 66 is added to a number, 15 is the result. What is the number?

    a. -51

    b. -81

    c. 51

    d. 81

680. Which of the following is equivalent to $4n^2(5np + p^2)$?

    a. $20n^2p + p^2$

    b. $20n^3p + p^2$

    c. $20n^3p + 4p^2$

    d. $20n^3p + 4n^2p^2$

681. Simplify the expression: $\frac{x^2 + 2x - 15}{x + 5}$.

    a. $x + 5$

    b. $x + 3$

    c. $x - 5$

    d. $x - 3$

682. Which of the following is equivalent to $2y^2$?

    a. $2(y + y)$

    b. $2y(y)$

    c. $y^2 + 2$

    d. $y + y + y + y$

683. $x(3x^2 + y) =$

    a. $4x^2 + xy$

    b. $4x^2 + x + y$

    c. $3x^3 + 2xy$

    d. $3x^3 + xy$

684. Which of the following is equivalent to the product of the expressions $(3x^2y)$ and $(2xy^2)$?

a. $5x^2y^2$

b. $5x^3y^3$

c. $6x^2y^2$

d. $6x^3y^3$

685. An equation of the form $\frac{a}{b} = \frac{c}{d}$ is

a. an inequality.

b. a variable.

c. a proportion.

d. a monomial.

686. Which of the following expressions best represents the sum of two numbers, a and b, divided by a third number, c?

a. $a + b \div c$

b. $(a + b) \div c$

c. $a \div (b + c)$

d. $a \div b + c$

687. If $\frac{x}{2} + \frac{x}{6} = 4$, what is x?

a. $\frac{1}{24}$

b. $\frac{1}{6}$

c. 3

d. 6

688. If $\frac{2}{5} = \frac{x}{45}$, what is x?

a. 9

b. 12

c. 18

d. 90

▶ **Set 44** (Answers begin on page 197.)

689. Solve for x in the following equation:
$\frac{1}{5}x - 6 = 3$
   a. 9
   b. 25
   c. 35
   d. 45

690. Solve the following equation for y: $8y - 3 = 1$
   a. $-\frac{3}{2}$
   b. $\frac{3}{2}$
   c. $\frac{1}{2}$
   d. $-\frac{1}{2}$

691. What is the value of y when $x = 5$ and
$y = 2(x^2 + 10) - 3$
   a. 61
   b. 63
   c. 65
   d. 67

692. If $\frac{1}{19} = \frac{x}{76}$, what is x?
   a. 3
   b. 3.5
   c. 4
   d. 5

693. Solve for p in the following equation:
$2.5p + 6 = 18.5$
   a. 5
   b. 10
   c. 15
   d. 20

694. Which of the following is the largest possible solution to the following inequality:
$\frac{1}{3}x - 3 \leq 5$
   a. $\frac{2}{3}$
   b. $\frac{8}{3}$
   c. 6
   d. 24

695. Which of the following linear equations has a slope of -2 and a y-intercept of 3?
   a. $y = 2x + 3$
   b. $y = 3x - 2$
   c. $y = -2x + 3$
   d. $y = \frac{1}{2}x - 3$

696. Evaluate the following expression if $a = 3$, $b = 4$, and $c = -2$: $(ab - ac) \div abc$.
   a. $-\frac{7}{8}$
   b. $-\frac{3}{4}$
   c. $-\frac{1}{4}$
   d. $\frac{1}{4}$

697. If $(\frac{x}{5} - \frac{x}{10} = 4)$, what is x?
   a. 4
   b. 20
   c. 40
   d. 80

698. A line passes through the points $(0,-1)$ and $(2,3)$. What is the equation for the line?
   a. $y = \frac{1}{2}x - 1$
   b. $y = \frac{1}{2}x + 1$
   c. $y = 2x - 1$
   d. $y = 2x + 1$

699. When the product of three and a number is taken away from the sum of that number and six, the result is zero. What is the number?

    a. 3

    b. 7

    c. 9

    d. 14

700. What is the slope of the linear equation $3y - x = 9$?

    a. $\frac{1}{3}$

    b. -3

    c. 3

    d. 9

701. The tens digit is four times the ones digit in a certain number. If the sum of the digits is 10, what is the number?

    a. 93

    b. 82

    c. 41

    d. 28

702. Which expression best describes the sum of three numbers multiplied by the sum of their reciprocals?

    a. $(a + b + c)(\frac{1}{a} + \frac{1}{b} + \frac{1}{c})$

    b. $(a)(\frac{1}{a}) + (b)(\frac{1}{b}) + (c)(\frac{1}{c})$

    c. $(a + b + c) \div (\frac{1}{a})(\frac{1}{b})(\frac{1}{c})$

    d. $(a)(b)(c) + (\frac{1}{a})(\frac{1}{b})(\frac{1}{c})$

703. Find the sum of $4x - 7y$ and $7x + 7y$.

    a. $11x$

    b. $14y$

    c. $11x + 14y$

    d. $11x - 14y$

704. Eighty-eight is the result when one-half of the sum of 24 and a number is all taken away from three times the number. What is the number?

    a. 16

    b. 40

    c. 56

    d. 112

▶**Set 45** (Answers begin on page 198.)

705. Which of the following lines have positive slopes?

a.

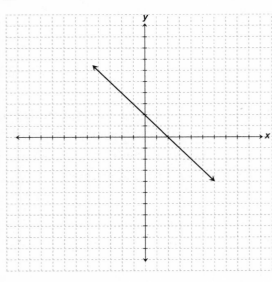

b.

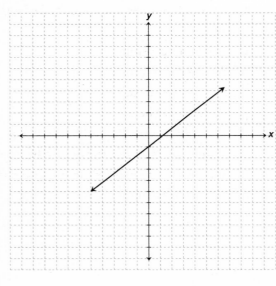

c.

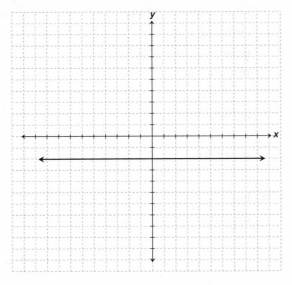

d.

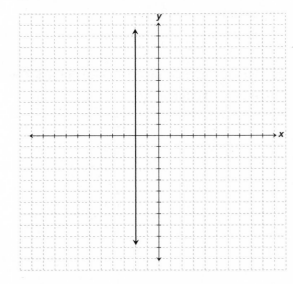

706. Generic oatmeal costs $\frac{2}{3}$ the price of the leading name brand. If the generic brand is $1.50, how much does the name brand cost?
  a. $1.00
  b. $1.75
  c. $2.00
  d. $2.25

707. A traveling circus can sell 250 admission tickets for $8 each. But if the tickets cost $6 each the circus can sell 400 tickets. How much larger are ticket sales when they cost $6 each than when they cost $8 each?

   a. $160
   b. $400
   c. $500
   d. $1,700

708. Forty-eight dollars in tips is to be divided among three restaurant waiters. Twila gets three times more than Jenny, and Betty receives four times as much as Jenny. How much does Betty receive?

   a. $18
   b. $16
   c. $24
   d. $6

709. How many solutions are there to the following system of equations?

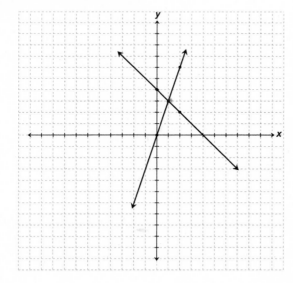

   a. 0
   b. 1
   c. 2
   d. infinite

710. A family of three ate dinner at a restaurant, with a total bill of $48. If the mother's meal cost $\frac{5}{4}$ as much as the father's, and the child's meal was $\frac{3}{4}$ that of the father's, how much was the father's meal?

   a. $12
   b. $14
   c. $16
   d. $20

711. Six pounds of a dried fruit mixture that costs $3 per pound and $1\frac{1}{2}$ pounds of nuts costing $7 per pound are mixed together. What is the cost per pound of this mixture?

   a. $1.33
   b. $3.80
   c. $5.25
   d. $8.75

712. A helicopter flies over a river at 6:02 A.M. and arrives at a heliport 20 miles away at 6:17 A.M. How many miles per hour was the helicopter traveling?

   a. 120 miles per hour
   b. 300 miles per hour
   c. 30 miles per hour
   d. 80 miles per hour

713. Two saline solutions are mixed. Twelve liters of 5% solution are mixed with four liters of 4% solution. What percent saline is the final solution?

   a. 4.25%
   b. 4.5%
   c. 4.75%
   d. 5%

714. D'Andre rides the first half of a bike race in two hours. If his partner, Adam, rides the return trip five miles per hour less, and it takes him three hours, how fast was D'Andre traveling?

a. 10 miles per hour

b. 15 miles per hour

c. 20 miles per hour

d. 25 miles per hour

715. Lee can catch 10 fish in an hour, and Charles can catch five fish in two hours. How long will Charles have to fish in order to catch the same number of fish that Lee would catch in two hours?

a. 2 hours

b. 4 hours

c. 6 hours

d. 8 hours

716. A grain elevator operator wants to mix two batches of corn with a resultant mix of 54 pounds per bushel. If he uses 20 bushels of 56 pounds per bushel corn, which of the following expressions gives the amount of 50 pounds per bushel corn needed?

a. $56x + 50x = 2x \times 54$

b. $20 \times 56 + 50x = (x + 20) \times 54$

c. $20 \times 56 + 50x = 2x \times 54$

d. $56x + 50x = (x + 20) \times 54$

717. Jared and Linda are both salespeople at a certain electronics store. If they made 36 sales one day, and Linda sold three less than twice Jared's sales total, how many units did Jared sell?

a. 19

b. 15

c. 12

d. 13

718. It will take John four days to string a certain fence. If Mary could string the same fence in three days, how long will it take them if they work together?

a. $3\frac{1}{2}$ days

b. 3 days

c. $2\frac{2}{7}$ days

d. $1\frac{5}{7}$ days

719. A recipe serves four people and calls for $1\frac{1}{2}$ cups of broth. If you want to serve six people, how much broth do you need?

a. 2 cups

b. $2\frac{1}{4}$ cups

c. $2\frac{1}{3}$ cups

d. $2\frac{1}{2}$ cups

720. If $x - 1$ represents an odd integer, which of the following represents the next larger odd integer?

a. $x - 3$

b. x

c. $x + 1$

d. $x + 2$

▶ **Set 46** (Answers begin on page 199.)

721. Which of the following is equivalent to $(x-3)(x+7)$?
   a. $x^2 - 3x - 21$
   b. $x^2 - 4x - 21$
   c. $x^2 + 4x - 21$
   d. $x^2 - 21$

722. Karl is four times as old as Pam, who is one-third as old as Jackie. If Jackie is 18, what is the sum of their ages?
   a. 64
   b. 54
   c. 48
   d. 24

723. Solve for x in terms of r and s: $s = 2x - r$.
   a. $x = s + r - 2$
   b. $x = 2s - r$
   c. $x = \frac{s+r}{2}$
   d. $x = \frac{2}{s+r}$

724. Three coolers of water per game are needed for a baseball team of 25 players. If the roster is expanded to 40 players, how many coolers are needed?
   a. 4
   b. 5
   c. 6
   d. 7

725. A pancake recipe calls for $1\frac{1}{2}$ cups of flour in order to make 14 pancakes. If $2\frac{1}{4}$ cups of flour are used, how many pancakes can be made?
   a. 18
   b. 21
   c. 24
   d. 27

726. The perimeter of a triangle is 25 inches. If side a is twice side b, which is $\frac{1}{2}$ side c, what is the length of side b?
   a. 5 inches
   b. 8 inches
   c. 10 inches
   d. 15 inches

727. How many gallons of a solution that is 75% antifreeze must be mixed with four gallons of a 30% solution to obtain a mixture that is 50% antifreeze?
   a. 2 gallons
   b. 3 gallons
   c. 3.2 gallons
   d. 4 gallons

728. Choose the equation that represents the graph in the following figure.

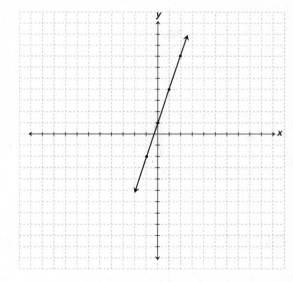

   a. $y = 3x + 1$
   b. $y = -3x - 1$
   c. $y = 3x - 1$
   d. $y = \frac{1}{3}x + 1$

729. A shopper can spend no more than $2 per pound on fruit and wants seven pounds of bananas at a cost of $0.50 per pound. How many pounds of raspberries can he buy if raspberries cost $4 per pound?

a. 5.25 pounds

b. 0.5 pound

c. 5.75 pounds

d. 2 pounds

730. A television costs $400 to purchase. Renting the same television requires making a $50 non-refundable deposit plus monthly payments of $25. After how many months will the cost to rent the television equal the cost to purchase it?

a. 8

b. 14

c. 16

d. 24

731. Eric and Margaret make $1,460 together one week. Eric makes $20 per hour, and Margaret makes $25 per hour. If Eric worked 10 hours more than Margaret one week, how many hours did Margaret work?

a. 25

b. 28

c. 33

d. 38

732. Multiply the binomials: $(3x + 4)(x - 6)$.

a. $3x^2 - 22x - 24$

b. $3x^2 + 14x - 24$

c. $3x^2 - 14x - 24$

d. $3x^2 + 14x + 24$

733. How much simple interest is earned on $767 if it is deposited in a bank account paying an annual interest rate of $7\frac{1}{8}$% interest for nine months? (*Interest = principal × rate × time*, or $I = PRT$)

a. $20.56

b. $64.13

c. $40.99

d. $491.83

734. A neighbor has three dogs. Fluffy is half the age of Muffy, who is one-third as old as Spot, who is half the neighbor's age, which is 24. How old is Fluffy?

a. 2

b. 4

c. 6

d. 12

735. A piggy bank contains $8.20 in coins. If there are an equal number of quarters, nickels, dimes, and pennies, how many of each denomination are there?

a. 10

b. 20

c. 30

d. 40

736. Simplify: $3(6x^4)^2$.

a. $18x^6$

b. $18x^8$

c. $108x^6$

d. $108x^8$

▶ **Set 47** (Answers begin on page 200.)

737. If Samantha deposits $385 today into a savings account paying 4.85% simple interest annually, how much interest will accrue in one year? (*Interest = principal × rate × time*, or $I = PRT$)

a. $1.86
b. $18.67
c. $186.73
d. $1,867.24

738. Veronica took a trip to the lake. If she drove steadily for five hours traveling 220 miles, what was her average speed for the trip?

a. 44 miles per hour
b. 55 miles per hour
c. 60 miles per hour
d. 66 miler per hour

739. Factor the expression completely: $x^2 - 25$.

a. $x(x - 25)$
b. $(x + 5)(x - 5)$
c. $(x + 5)(x + 5)$
d. $(x - 5)(x - 5)$

740. Factor the expression completely: $x^2 - 2x - 48$.

a. $(x + 8)(x - 6)$
b. $(x + 8)(x + 6)$
c. $(x - 8)(x + 6)$
d. $(x - 8)(x - 6)$

741. Match the inequality with the graph.

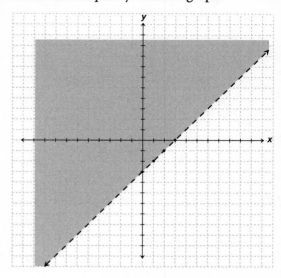

a. $y < x - 3$
b. $y > x - 3$
c. $y \geq x - 3$
d. $y \leq x - 3$

742. Which of the following points is the solution to the system of equations?
$$y = -x + 10$$
$$y = x - 2$$

a. $(2,10)$
b. $(2,0)$
c. $(3,6)$
d. $(6,4)$

743. If jogging for one mile uses 150 calories and brisk walking for one mile uses 100 calories, a jogger has to go how many times as far as a walker to use the same number of calories?

a. $\frac{1}{2}$
b. $\frac{2}{3}$
c. $\frac{3}{2}$
d. 2

744. How much water must be added to one liter of a 5% saline solution to get a 2% saline solution?

a. 1 liter

b. 1.5 liters

c. 2 liters

d. 2.5 liters

745. Joan will be twice Tom's age in three years when Tom will be 40. How many years old is Joan now?

a. 20

b. 80

c. 77

d. 37

746. Multiply: $\frac{3xy^2}{4y} \cdot \frac{y^3}{9x}$.

a. $\frac{3x}{4y}$

b. $\frac{3x^2}{4y}$

c. $\frac{xy^2}{4}$

d. $\frac{xy^2}{12}$

747. Water is coming into a tank three times as fast as it is going out. After one hour, the tank contains 11,400 gallons of water. How fast is the water coming in?

a. 3,800 gallons per hour

b. 5,700 gallons per hour

c. 11,400 gallons per hour

d. 17,100 gallons per hour

748. Jim is twice as old as Sally, who is half as old as Bill. The sum of their ages is 60. How old is Jim?

a. 24

b. 26

c. 30

d. 32

749. A man fishing on a riverbank sees a boat pass by. The man estimates the boat is traveling 20 miles per hour. If his estimate is correct, how many minutes will it be before he sees the boat disappear around a bend in the river half a mile away?

a. 14 minutes

b. 1.4 minutes

c. 2.4 minutes

d. 1.5 minutes

750. After three days, some hikers discover that they have used $\frac{2}{5}$ of their supplies. At this rate, how many more days can they go forward before they have to turn around?

a. 0.75 day

b. 1.5 days

c. 3.75 days

d. 4.5 days

751. Solve for all values of x in the equation: $x^2 - 25 = 0$.

a. 5

b. 0, 5

c. −5

d. 5, −5

752. Jason's hair salon charges $63 for a haircut and color, which is $\frac{3}{4}$ of what Lisa's hair salon charges. How much does Lisa's hair salon charge?

a. $65

b. $21

c. $42

d. $84

▶ **Set 48** (Answers begin on page 202.)

753. Divide: $\frac{6a^2b}{2c} \div \frac{ab^2}{4c^4}$.
   a. $\frac{24ac}{b}$
   b. $\frac{12ac^3}{b}$
   c. $\frac{24ac^3}{b}$
   d. $12abc^3$

754. Five oranges, when removed from a basket containing three more than seven times as many oranges, leaves how many in the basket?
   a. 21
   b. 28
   c. 33
   d. 38

755. How many ounces of candy costing $1 per ounce must be mixed with 6 ounces of candy costing $0.70 per ounce to yield a mixture costing $0.80 per ounce?
   a. $\frac{6}{7}$ ounce
   b. 3 ounces
   c. 9 ounces
   d. $20\frac{7}{10}$ ounces

756. Dave can wash and wax a car in four hours. Mark can do the same job in three hours. If both work together, how long will it take to wash and wax a car?
   a. 1.7 hours
   b. 2.3 hours
   c. 3 hours
   d. 3.3 hours

757. How long will it take Sheri to walk to a store five miles away if she walks at a steady pace of three miles per hour?
   a. 0.60 hour
   b. 1.67 hours
   c. 3.33 hours
   d. 15 hours

758. Jeff was 10 minutes early for class. Dee came in four minutes after Mae, who was half as early as Jeff. How many minutes early was Dee?
   a. 1 minute
   b. 2 minutes
   c. 2.5 minutes
   d. 6 minutes

759. Jan loaned Ralph $45 expecting him to repay $50 in one month. What is the amount of annual simple interest on this loan? (R = the annual rate of simple interest. *Interest = principal × rate × time*, or $I = PRT$)
   a. 5%
   b. 33%
   c. 60%
   d. 133%

760. Twelve people entered a room. Three more than two-thirds of these people then left. How many people remain in the room?
   a. 0
   b. 1
   c. 2
   d. 7

761. A 24-inch-tall picture is 20% as tall as the ceiling is high. How high is the ceiling?

   a. 4.8 feet

   b. 10 feet

   c. 12 feet

   d. 120 feet

762. Belinda is building a garden shed. When she helped her neighbor build an identical shed, it took them both 22 hours to complete the job. If it would have taken her neighbor, working alone, 38 hours to build the shed, how long will it take Belinda, working alone, to build her shed?

   a. 33.75 hours

   b. 41.00 hours

   c. 41.25 hours

   d. 52.25 hours

763. If nine candles are blown out on a birthday cake containing seven times as many candles altogether, how many candles are there in all?

   a. 2

   b. 16

   c. 63

   d. 72

764. Simplify the radical: $\sqrt{\frac{81x^2}{y^2}}$.

   a. $\frac{9x}{y}$

   b. $9x$

   c. $9xy$

   d. $\frac{9x^2}{y}$

765. Which of the following linear equations has a negative slope?

   a. $6 = y - x$

   b. $y = 4x - 5$

   c. $-5x + y = 1$

   d. $6y + x = 7$

766. Rosa finds the average of her three most recent golf scores by using the following expression, where a, b, and c are the three scores: $\frac{a+b+c}{3} \times 100$. Which of the following would also determine the average of her scores?

   a. $(\frac{a}{3} + \frac{b}{3} + \frac{c}{3}) \times 100$

   b. $\frac{\frac{a+b+c}{3}}{100}$

   c. $\frac{(a+b+c) \times 3}{100}$

   d. $\frac{a \times b \times c}{3} + 100$

767. Simplify the radical completely: $\sqrt{64x^5y^8}$.

   a. $8x^2y^4\sqrt{x}$

   b. $8x^4y^8\sqrt{x}$

   c. $64x^2y^4$

   d. $8x^5y^8$

768. While driving home from work, Sally runs over a nail, causing a tire to start leaking. She estimates that her tire is leaking one pound per square inch (psi) every 20 seconds. Assuming that her tire leaks at a constant rate and her initial tire pressure was 36 psi, how long will it take her tire to completely deflate?

   a. 1.8 minutes

   b. 3.6 minutes

   c. 12 minutes

   d. 18 minutes

▶ **Set 49** (Answers begin on page 203.)

769. Mark needs to work five-sixths of a year to pay off his car loan. If Mark begins working on March 1, at the end of what month will he first be able to pay off his loan?
a. June
b. August
c. October
d. December

770. Solve the equation for b: $\sqrt{b-4} = 5$.
a. 1
b. 9
c. 21
d. 29

771. Find the sum: $\frac{2w}{z} + \frac{5w}{z}$.
a. $\frac{7w}{2z}$
b. $\frac{7w}{z^2}$
c. $\frac{7w}{z}$
d. $7w$

772. Some birds are sitting in an oak tree. Ten more birds land. More birds arrive until there are a total of four times as many birds as the oak tree had after the ten landed. A nearby maple tree has 16 fewer than 12 times as many birds as the oak tree had after the 10 landed. If both trees now have the same number of birds, how many birds were originally in the oak tree before the first 10 landed?
a. 4
b. 7
c. 16
d. 24

773. A jar of coins totaling $4.58 contains 13 quarters and five nickels. There are twice as many pennies as there are dimes. How many dimes are there?
a. 5
b. 9
c. 18
d. 36

774. If a function is defined as $f(x) = 5x^2 - 3$, what is the value of $f(-2)$?
a. −13
b. 7
c. −23
d. 17

775. A hiker walks from his car to a distant lake and back again. He walks on smooth terrain for two hours until he reaches a five-mile-long rocky trail. His pace along the five-mile-long trail is two miles per hour. If he walks steadily with no stops, how long will it take the hiker to complete the entire trip from his car to the lake and back again?
a. 4.5 hours
b. 9 hours
c. 12 hours
d. 16 hours

776. The total pressure of a mix of gases in a container is equal to the sum of the partial pressures of each of the gases in the mixture. A mixture contains nitrogen, oxygen, and argon, and the pressure of nitrogen is twice the pressure of oxygen, which is three times the pressure of argon. If the partial pressure of nitrogen is four pounds per square inch (psi), what is the total pressure?

a. 24 psi

b. 16 psi

c. $14\frac{2}{3}$ psi

d. $6\frac{2}{3}$ psi

777. Suppose the amount of radiation that could be received from a microwave oven varies inversely as the square of the distance from it. How many feet away must you stand to reduce your potential radiation exposure to $\frac{1}{16}$ the amount you could receive standing one foot away?

a. 16 feet

b. 4 feet

c. 32 feet

d. 8 feet

778. Which of the following linear equations has an undefined slope?

a.

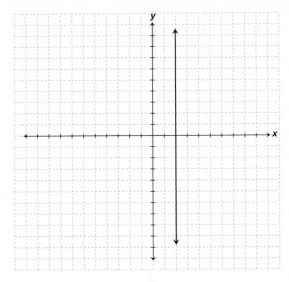

b.

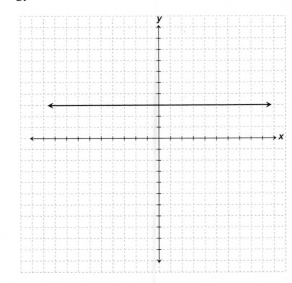

c.

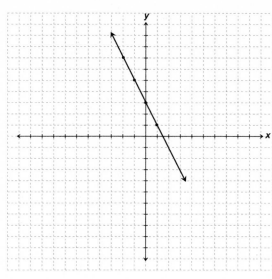

d.

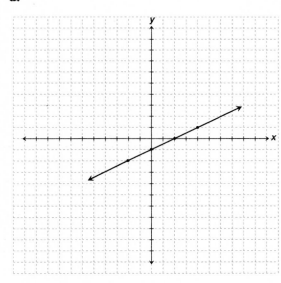

779. Solve the equation for a: $\sqrt{2a + 6} - 4 = 6$.

a. 2

b. 17

c. 23

d. 47

780. If Willie must pay an employment agency his first month's salary as a placement fee, how much of his $28,000 first year's salary will Willie end up with?

a. $2,333

b. $3,360

c. $24,343

d. $25,667

781. Jack went hiking in bad weather wearing a backpack filled with 60 pounds of supplies. A couple of miles into his hike, he became tired and discarded supplies equal to $\frac{1}{3}$ of the 60 pounds. A few miles later it started to snow, and he discarded another $\frac{2}{5}$ of the original 60 pounds. How much did Jack discard altogether during his hike?

a. 5 pounds

b. 10 pounds

c. 20 pounds

d. 44 pounds

782. If a school buys three computers at a, b, and c dollars each, and the school pays only 90% because of a discount, which expression would determine the average price per computer paid by the school?

a. $\frac{0.9 \times (a + b + c)}{3}$

b. $\frac{(a + b + c)}{0.9}$

c. $(a + b + c) \times 0.9$

d. $\frac{(a + b + c)}{3}$

783. Three apples and twice as many oranges add up to one-half the number of cherries in a fruit basket. How many cherries are there?

   a. 11
   b. 18
   c. 21
   d. 24

784. Rudy forgot to replace his gas cap the last time he filled up his car with gas. The gas is evaporating out of his 14-gallon tank at a constant rate of $\frac{1}{3}$ gallon per day. How much gas does Rudy lose in one week?

   a. 2 gallons
   b. $2\frac{1}{3}$ gallons
   c. $4\frac{2}{3}$ gallons
   d. 6 gallons

▶ **Set 50** (Answers begin on page 204.)

785. Which ratio best expresses the following: Five hours is what percent of a day?

a. $\frac{5}{100} = \frac{x}{24}$

b. $\frac{5}{24} = \frac{24}{x}$

c. $\frac{5}{24} = \frac{x}{100}$

d. $\frac{x}{100} = \frac{24}{5}$

786. White flour and whole wheat flour are mixed together in a ratio of five parts white flour to one part whole wheat flour. How many pounds of white flour are in 48 pounds of this mixture?

a. 8 pounds

b. 9.6 pounds

c. 40 pounds

d. 42 pounds

787. Timmy can sell 20 glasses of lemonade for 10 cents per glass. If he raises the price to 25 cents per glass, Timmy estimates he can sell seven glasses. If so, how much more money will Timmy make by charging 25 cents instead of 10 cents per glass?

a. −$0.25

b. $0

c. $0.10

d. $0.50

788. Bill lives five miles away from school. Tammy lives half as far away from school. Joe's distance from school is halfway between Bill and Tammy's. How far away from school does Joe live?

a. 1.25 miles

b. 3.75 miles

c. 6.25 miles

d. 7.5 miles

789. Pamela's annual salary is six times as much as Adrienne's, who earns five times more than Beverly, who earns $4,000. If Valerie earns one-half as much as Pam, what is Valerie's annual salary?

a. $25,000

b. $60,000

c. $90,000

d. $110,000

790. Yolanda and Gertrude are sisters. When one-fourth of Gertrude's age is taken away from Yolanda's age, the result is twice Gertrude's age. If Yolanda is nine, how old is Gertrude?

a. 2.25 years old

b. 4 years old

c. 4.5 years old

d. 18 years old

791. How many pounds of chocolates costing $5.95 per pound must be mixed with three pounds of caramels costing $2.95 per pound to obtain a mixture that costs $3.95 per pound?

a. 1.5 pounds

b. 3 pounds

c. 4.5 pounds

d. 8 pounds

792. Kathy charges $7.50 per hour to mow a lawn. Sharon charges 1.5 times as much to do the same job. How much does Sharon charge to mow a lawn?

a. $5.00 per hour

b. $11.25 per hour

c. $10.00 per hour

d. $9.00 per hour

793. Laura saves at three times the rate Hazel does. If it takes Laura $1\frac{1}{2}$ years to save \$1,000, how many years will it take Hazel to save this amount?

   a. 1

   b. 3.5

   c. 4.5

   d. 6

794. Find the difference: $\frac{9}{2a} - \frac{3w}{6a^3}$.

   a. $\frac{9a^2 - w}{2a^3}$

   b. $\frac{9 - 3w}{6a^3}$

   c. $\frac{9a^2 - w}{6a^3}$

   d. $\frac{27a^2 - w}{2a^3}$

795. If a function is defined as $g(x) = x^2 - 4x + 1$, what is the value of $g(1)$?

   a. -3

   b. -2

   c. -1

   d. 1

796. Fire departments commonly use the following formula to find out how far from a wall to place the base of a ladder: (Length of ladder ÷ 5 feet) + 2 feet = distance from the wall. Using this formula, if the base of a ladder is placed 10 feet from a wall, how tall is the ladder?

   a. 40 feet

   b. 48 feet

   c. 72 feet

   d. 100 feet

797. William has one-third as many toys as Tammy, who has four times more toys than Edgar, who has six. How many toys does Jane have if she has five more than William?

   a. 4

   b. 13

   c. 8

   d. 9

798. Susan and Janice are sign painters. Susan can paint a sign in six hours, while Janice can paint the same sign in five hours. If both worked together, how long would it take them to paint a sign?

   a. 2.53 hours

   b. 2.73 hours

   c. 3.00 hours

   d. 3.23 hours

799. Match the graph with the inequality: $5y \geq 10x - 15$.

a.

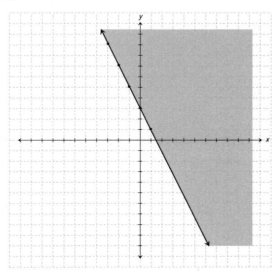

b.

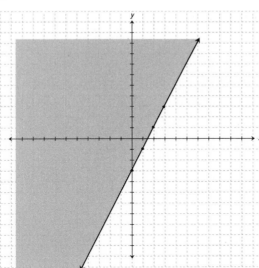

c.

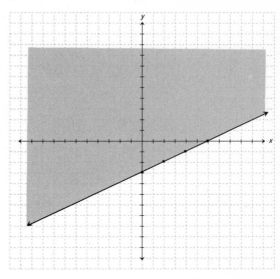

d.

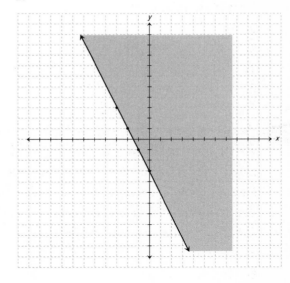

800. Solve the following system of equations algebraically by substitution:

$$y + 2x = 3$$
$$x = 3y + 5$$

a. $(2,-1)$

b. $(2,0)$

c. $(-1,-2)$

d. $(2,-2)$

▶ **Set 51** (Answers begin on page 205.)

801. Which of the following systems contains two parallel lines?

   a. $x = 5$
       $y = 5$

   b. $y = -x$
       $y = x - 1$

   c. $x - y = 7$
       $2 - y = -x$

   d. $y = 3x + 4$
       $2x + 4 = y$

802. How much simple interest is earned on $300 deposited for 30 months in a savings account paying $7\frac{3}{4}\%$ simple interest annually? *(Interest = principal × rate × time, or I = PRT)*

   a. $5.81
   b. $19.76
   c. $23.25
   d. $58.13

803. Jason is six times as old as Kate. In two years, Jason will be twice as old as Kate is then. How old is Jason now?

   a. 6 months old
   b. 3 years old
   c. 6 years old
   d. 9 years old

804. Mike types three times as fast as Nick. Together they type 24 pages per hour. If Nick learns to type as fast as Mike, how many pages will they be able to type per hour?

   a. 18
   b. 30
   c. 36
   d. 40

805. Thirteen percent of a certain school's students are A students, 15% are C students, and 20% make mostly Ds. If 16% of the students are B students, what percent are failing?

   a. 25%
   b. 36%
   c. 48%
   d. 64%

806. Which equation is represented by the following graph?

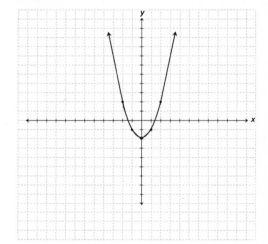

   a. $y = x^2 - 4$
   b. $y = x^2 - 2$
   c. $y = x^2$
   d. $y = x^2 + 4$

807. A four-pound mixture of raisins and nuts is $\frac{2}{3}$ raisins. How many pounds of nuts are there?

   a. 1.3 pounds
   b. 1.6 pounds
   c. 2.4 pounds
   d. 2.6 pounds

808. When estimating distances from aerial photographs, the formula is $(A \times l) \div f = D$, where A is the altitude, l is the measured length on the photograph, f is the focal length of the camera, and D is the actual distance. If the focal length is six inches, the measured length across a lake on a photograph is three inches, and the lake is 3,000 feet across, at what altitude was the photograph taken?

a. 1,500 feet

b. 3,000 feet

c. 4,500 feet

d. 6,000 feet

809. Solve for x: $\frac{2}{x-5} + \frac{3}{x+3} = \frac{11}{x^2 - 2x - 15}$.

a. 3

b. 4

c. 6

d. 11

810. A new telephone directory with 596 pages has 114 less than twice as many pages as last year's edition. How many pages were in last year's directory?

a. 298

b. 355

c. 412

d. 482

811. A vacationing family travels 300 miles on their first day. The second day they travel only $\frac{2}{3}$ as far. But on the third day they are able to travel $\frac{3}{4}$ as many miles as the first two days combined. How many miles have they traveled altogether?

a. 375 miles

b. 875 miles

c. 525 miles

d. 500 miles

812. A rain barrel contained four gallons of water just before a thunderstorm. It rained steadily for eight hours, filling the barrel at a rate of six gallons per day. How many gallons of water did the barrel have after the thunderstorm?

a. 4 gallons

b. 6 gallons

c. 7 gallons

d. 9 gallons

813. Find three consecutive odd integers whose sum is 117.

a. 39, 39, 39

b. 38, 39, 40

c. 37, 39, 41

d. 39, 41, 43

814. Three species of songbirds live in a certain plot of trees, totaling 120 birds. If species A has three times as many birds as species B, which has half as many birds as C, how many birds belong to species C?

a. 20

b. 30

c. 40

d. 60

815. The base of a rectangle is seven times the height. If the perimeter is 32 meters, what is the area?

a. 28 square meters

b. 24 square meters

c. 16 square meters

d. 14 square meters

816. Which equation is represented by the following graph?

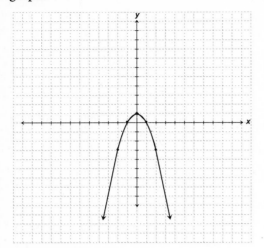

   a. $y = -x^2 - 1$

   b. $y = -x^2$

   c. $y = -x^2 + 1$

   d. $y = x^2 + 1$

6 ▶ Geometry

The 12 sets of basic geometry problems in this section involve lines, angles, triangles, rectangles, squares, and circles. For example, you may be asked to find the area or perimeter of a shape, the length of a line, or the circumference of a circle. In addition, the word problems will illustrate how closely geometry is related to the real world and to everyday life.

▶ **Set 52** (Answers begin on page 207.)

817. Each of the following figures has exactly two pairs of parallel sides EXCEPT a
   a. parallelogram.
   b. rhombus.
   c. trapezoid.
   d. square.

818. How many faces does a cube have?
   a. 4
   b. 6
   c. 8
   d. 12

819. A polygon is a plane figure composed of connected lines. How many connected lines must there be to make a polygon?
   a. 3 or more
   b. 4 or more
   c. 5 or more
   d. 6 or more

820. An acute angle is
   a. 180°.
   b. greater than 90°.
   c. 90°.
   d. less than 90°.

821. A straight angle is
   a. exactly 180°.
   b. between 90° and 180°.
   c. 90°.
   d. less than 90°.

822. A right angle is
   a. 180°.
   b. greater than 90°.
   c. exactly 90°.
   d. less than 90°.

823. Which of the following statements is true?
   a. Parallel lines intersect at right angles.
   b. Parallel lines never intersect.
   c. Perpendicular lines never intersect.
   d. Intersecting lines have two points in common.

824. A triangle has two congruent sides, and the measure of one angle is 40°. Which of the following types of triangles is it?
   a. isosceles
   b. equilateral
   c. right
   d. scalene

825. A triangle has one 30° angle and one 60° angle. Which of the following types of triangles is it?
   a. isosceles
   b. equilateral
   c. right
   d. scalene

826. A triangle has angles of 71° and 62°. Which of the following best describes the triangle?
   a. acute scalene
   b. obtuse scalene
   c. acute isosceles
   d. obtuse isosceles

827. In which of the following are the diagonals of the figure always congruent and perpendicular?
  a. isosceles trapezoid
  b. square
  c. isosceles triangle
  d. rhombus

828. If one angle of a triangle measures 42° and the second measures 59°, what does the third angle measure?
  a. 101°
  b. 89°
  c. 90°
  d. 79°

829. Which of the following could describe a quadrilateral with two pairs of parallel sides and two interior angles that measure 65°?
  a. square
  b. triangle
  c. rectangle
  d. rhombus

830. If pentagon *ABCDE* is similar to pentagon *FGHIJ*, and *AB* = 10, *CD* = 5, and *FG* = 30, what is *HI*?
  a. $\frac{5}{3}$
  b. 5
  c. 15
  d. 30

831. What is the greatest area possible enclosed by a quadrilateral with a perimeter of 24 feet?
  a. 6 square feet
  b. 24 square feet
  c. 36 square feet
  d. 48 square feet

832. What is the difference in area between a square with a base of four feet and a circle with a diameter of four feet?
  a. $16 - 2\pi$ square feet
  b. $16 - 4\pi$ square feet
  c. $8\pi - 16$ square feet
  d. $16\pi - 16$ square feet

▶ **Set 53** (Answers begin on page 207.)

833. What is the difference in perimeter between a square with a base of four feet and a circle with a diameter of four feet?
- **a.** $8 - 2\pi$ square feet
- **b.** $16 - 2\pi$ square feet
- **c.** $16 - 4\pi$ square feet
- **d.** $16 - 8\pi$ square feet

834. The animal shelter is developing a new outdoor grass area for dogs. A fence needs to be purchased that will surround the entire grassy section. The dimensions of the area are 120 feet by 250 feet. How much fencing needs to be purchased?
- **a.** 740 feet
- **b.** 30,000 square feet
- **c.** 740 square feet
- **d.** 30,000 feet

835. What is the perimeter of the following triangle?

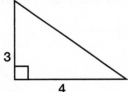

- **a.** 12
- **b.** 9
- **c.** 8
- **d.** 7

836. What is the perimeter of the following polygon?

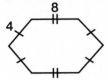

- **a.** 12
- **b.** 16
- **c.** 24
- **d.** 32

837. What is the perimeter of the following figure?

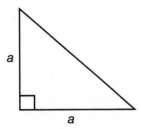

- **a.** $a^2 \div 2$
- **b.** $2a + 2a^2$
- **c.** $2a + \sqrt{2a^2}$
- **d.** $4a$

838. One side of a rectangle measures 834 centimeters and another measures 1,288 centimeters. What is the perimeter of the triangle?
- **a.** 2,148,384 square feet
- **b.** 1,074,192 square feet
- **c.** 4,244 feet
- **d.** 2,122 feet

839. A landlord is renovating an apartment. He is putting new tile in the rectangular kitchen. The length of the room is 10 feet and the width is 6.5 feet. What is the area that needs to be tiled?
- **a.** 650 square feet
- **b.** 16.5 square feet
- **c.** 65 square feet
- **d.** 33 square feet

840. What is the outer perimeter of the doorway shown here?

a. 12
b. 24
c. 20 + 2π
d. 24 + 2π

841. What is the perimeter of the following triangle?

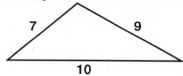

a. 90
b. 70
c. 26
d. 19

842. What is the perimeter of the following parallelogram?

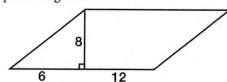

a. 26
b. 32
c. 48
d. 56

843. What is the perimeter of the shaded area, if the shape is a quarter circle with a radius of 8?

a. 2π
b. 4π
c. 2π + 16
d. 4π + 16

844. What is the perimeter of the following triangle?

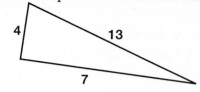

a. 11
b. 13
c. 24
d. 28

845. What is the length of the side labeled *x* of the following polygon if the perimeter is 40 inches?

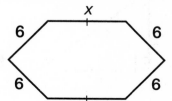

a. 6 inches
b. 8 inches
c. 10 inches
d. 16 inches

846. What is the perimeter of the following rectangle?

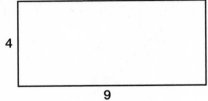

4

9

- **a.** 13
- **b.** 22
- **c.** 26
- **d.** 36

847. What lines must be parallel in the following diagram?

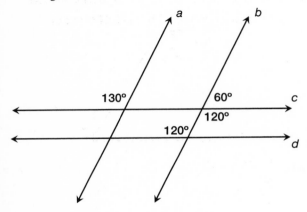

- **a.** *a* and *b*
- **b.** *a* and *d*
- **c.** *b* and *c*
- **d.** *c* and *d*

848. What is the measure of angle *F* in the following diagram?

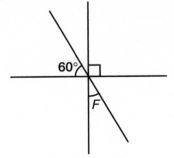

- **a.** 15°
- **b.** 45°
- **c.** 30°
- **d.** 90°

▶ **Set 54** (Answers begin on page 208.)

849. In the following figure, which pair of angles must be congruent?

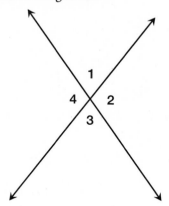

a. 1 and 2
b. 1 and 3
c. 3 and 4
d. 1 and 4

850. Which of the following lengths could form the sides of a triangle?
a. 1, 2, 3
b. 2, 2, 5
c. 2, 3, 6
d. 2, 3, 2

851. Which side of the following right triangle is the shortest, if angle *ACB* is 46°?

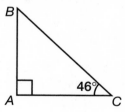

a. *AB*
b. *AC*
c. *BC*
d. All sides are equal.

852. Which side of the following triangle is the shortest, if angles *BAC* and *ABC* are 60°?

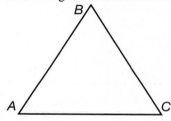

a. *AB*
b. *AC*
c. *BC*
d. All sides are equal.

853. Which of the angles in the following figure must be congruent if lines *l* and *m* are parallel?

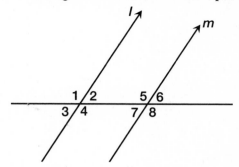

a. 1 and 2
b. 1 and 8
c. 2 and 8
d. 4 and 7

854. What is the measure of angle *ABC*?

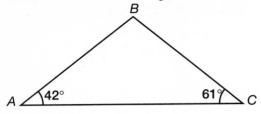

    a. 42°
    b. 61°
    c. 77°
    d. 103°

855. Plattville is 80 miles west and 60 miles north of Quincy. How long is a direct route from Plattville to Quincy?
    a. 100 miles
    b. 110 miles
    c. 120 miles
    d. 140 miles

856. Which is the longest side of the following triangle? (Note: not drawn to scale.)

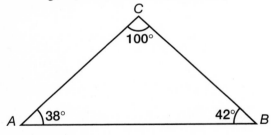

    a. *AB*
    b. *AC*
    c. *BC*
    d. *AC* and *BC*

857. What is the complement of a 42° angle?
    a. 48°
    b. 45°
    c. 138°
    d. 145°

858. Two angles are supplementary. One measures 84°. What does its supplement measure?
    a. 90°
    b. 276°
    c. 6°
    d. 96°

859. A line is drawn within a right angle. One angle that is formed by the line and the vertex of the right angle measures 32°. What does the other angle measure?
    a. 148°
    b. 58°
    c. 328°
    d. 45°

860. The area of a rectangular table is 72 square inches. The longer sides are 12 inches long. What is the width?
    a. 5 inches
    b. 6 inches
    c. 7 inches
    d. 8 inches

861. The area of a right triangle is 60 square centimeters. The height is 15 centimeters. How many centimeters is the base?
    a. 8 centimeters
    b. 4 centimeters
    c. 6 centimeters
    d. 12 centimeters

862. Which of these angle measures form a right triangle?
    a. 40°, 40°, 100°
    b. 20°, 30°, 130°
    c. 40°, 40°, 40°
    d. 40°, 50°, 90°

863. What is the measure of angle B in the following diagram?

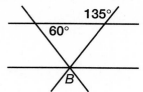

a. 45°

b. 60°

c. 75°

d. 130°

864. If the following figure is a regular decagon with a center at Q, what is the measure of the indicated angle?

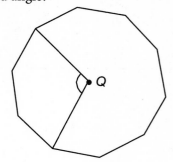

a. 45°

b. 80°

c. 90°

d. 108°

▶ **Set 55** (Answers begin on page 209.)

865. In the following figure, angle *POS* measures 90°. What is the measure of angle *ROQ*?

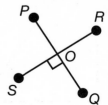

    a. 30°
    b. 45°
    c. 90°
    d. 180°

866. Triangles *RST* and *MNO* are similar. What is the length of line segment *MO*?

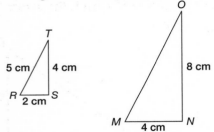

    a. 5 centimeters
    b. 10 centimeters
    c. 20 centimeters
    d. 32 centimeters

867. In the diagram, lines *a, b,* and *c* intersect at point *O*. Which of the following are NOT adjacent angles?

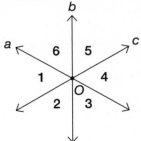

    a. ∠1 and ∠6
    b. ∠1 and ∠4
    c. ∠4 and ∠5
    d. ∠2 and ∠3

868. A square sandbox has sides that are 18 feet long. What is the perimeter of the sandbox?
    a. 342 feet
    b. 18 feet
    c. 324 feet
    d. 72 feet

869. Find the area of the following shape.

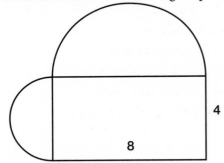

    a. 12 + 10π square units
    b. 32 + 10π square units
    c. 32 + 12π square units
    d. 12 + 12π square units

870. A triangle has sides that are consecutive even integers. The perimeter of the triangle is 24 inches. What is the length of the shortest side?
   a. 10 inches
   b. 8 inches
   c. 6 inches
   d. 4 inches

871. Assuming π is 3.14, what is the diameter of a circular pool with a circumference of 78 feet?
   a. 24.84 feet
   b. 12.42 feet
   c. 244.92 feet
   d. 122.46 feet

872. A garden is in the shape of a regular pentagon. The perimeter is 90 feet. What is the length of each side?
   a. 14 feet
   b. 16 feet
   c. 18 feet
   d. 20 feet

873. What is the area of a circle that has a diameter of 94 centimeters?
   a. 188π square centimeters
   b. 47π square centimeters
   c. 94π square centimeters
   d. $2{,}209\pi$ square centimeters

874. What is the perimeter of the figure shown?

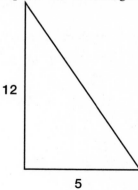

   a. 17
   b. 20
   c. 30
   d. 40

875. If the two triangles in the diagram are similar, with $\angle A$ equal to $\angle D$, what is the perimeter of triangle *DEF*?

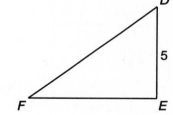
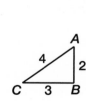

   a. 12
   b. 21
   c. 22.5
   d. 24.75

876. What is the measure of angle *C* in the following triangle?

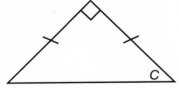

  a. 90°
  b. 45°
  c. 25°
  d. cannot be determined

877. What is the area of the following shaded triangle?

  a. 20 square units
  b. 25 square units
  c. 44 square units
  d. 46 square units

878. What is the measure of an interior angle of a regular hexagon?
  a. 60°
  b. 80°
  c. 120°
  d. 180°

879. Which side is the longest if triangle *A* is similar to triangle *B*?

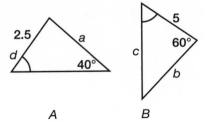

  a. *a*
  b. *b*
  c. *c*
  d. *d*

880. In the following diagram, if angle 1 is 30° and angle 2 is a right angle, what is the measure of angle 5?

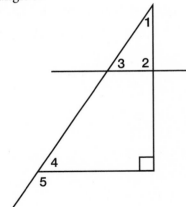

  a. 30°
  b. 60°
  c. 120°
  d. 140°

▶ **Set 56** (Answers begin on page 210.)

881. What is the measure of angle *A* in the following diagram?

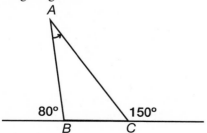

a. 40°
b. 50°
c. 60°
d. 70°

882. What is the measure of angle *ABC* if *ABCD* is a parallelogram, and the measure of angle *BAD* is 88°?

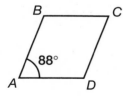

a. 88°
b. 90°
c. 92°
d. 102°

883. One base angle of an isosceles triangle is 70°. What is the vertex angle?

a. 130°
b. 90°
c. 70°
d. 40°

884. A circular fan is encased in a square guard. If one side of the guard is 12 inches, at what blade circumference will the fan just hit the guard?

a. 6 inches
b. 12 inches
c. 6π inches
d. 12π inches

885. If the circumference of a circle is half the area, what is the radius of the circle?

a. $\frac{1}{2}$
b. 2
c. 4
d. 8

886. The radius of a circle is 6.78 centimeters. What is the circumference?

a. 6.78π centimeters
b. 13.56π centimeters
c. 45.97π centimeters
d. 183.87π centimeters

887. What is the area of the following diagram?

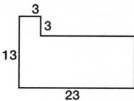

a. 239
b. 259
c. 299
d. 306

888. What is the perimeter of the following rectangle?

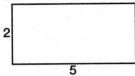

a. 7
b. 10
c. 14
d. 70

889. What is the volume of a pyramid that has a rectangular base five feet by three feet and a height of eight feet? Use $V = \frac{1}{3}(base)(height)$.

a. 16 cubic feet
b. 30 cubic feet
c. 40 cubic feet
d. 80 cubic feet

890. Two squares are similar. The areas of the smaller and larger squares are 81 square inches and 6,561 square inches respectively. What is the length of the sides of the larger square?

a. 9 inches
b. 81 inches
c. 1,640.25 inches
d. 6,561 inches

891. What is the length of a rectangle with an area of 420 square feet and a width of 20 feet?

a. 21 feet
b. 22 feet
c. 23 feet
d. 24 feet

892. All of the rooms on the top floor of a government building are rectangular, with eight-foot ceilings. One room is nine feet wide by 11 feet long. What is the combined area of the four walls, including doors and windows?

a. 99 square feet
b. 160 square feet
c. 320 square feet
d. 72 square feet

893. Jade made a painting on a rectangular canvas with the dimensions of 25 inches by 45 inches. How many square inches is the painting?

a. 280 square inches
b. 140 square inches
c. 1,125 square inches
d. 2,250 square inches

894. A square room has an area of 324 square feet. What is the perimeter?

a. 18 feet
b. 27 feet
c. 36 feet
d. 72 feet

895. Louise wants to wallpaper a room. It has one window that measures three feet by four feet, and one door that measures three feet by seven feet. The room is 12 feet by 12 feet, and is 10 feet tall. If only the walls are to be covered, and rolls of wallpaper are 100 square feet, what is the minimum number of rolls that she will need?

a. 4
b. 5
c. 6
d. 7

896. What is the area of the following triangle ?

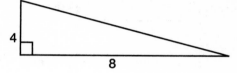

a. 6
b. 12
c. 16
d. 32

► **Set 57** (Answers begin on page 211.)

897. Find the area of the following parallelogram.

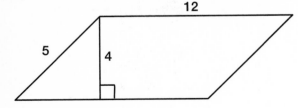

a. 48 square units
b. 68 square units
c. 72 square units
d. 240 square units

898. What is the area of the rectangle?

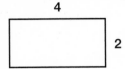

a. 6 square units
b. 8 square units
c. 12 square units
d. 16 square units

899. If the perimeter of a square mirror is 68 inches, what is the area in square inches?
a. 289 square inches
b. 17 square inches
c. 4,624 square inches
d. 272 square inches

900. The length of a rectangle is equal to four inches more than twice the width. Three times the length plus two times the width is equal to 28 inches. What is the area of the rectangle?
a. 8 square inches
b. 16 square inches
c. 24 square inches
d. 28 square inches

901. A rectangular box has a square base with an area of nine square feet. If the volume of the box is 36 cubic feet, what is the length of the longest object that can fit in the box? (Note: A calculator is needed for this problem.)
a. 3 feet
b. 5 feet
c. 5.8 feet
d. 17 feet

902. Dennis Sorensen is buying land on which he plans to build a cabin. He wants 200 feet in road frontage and a lot 500 feet deep. If the asking price is $9,000 an acre for the land, approximately how much will Dennis pay for his lot? (1 acre = 43,560 square feet)
a. $10,000
b. $20,700
c. $22,956
d. $24,104

903. In the following diagram, a circle of area 100π square inches is inscribed in a square. What is the length of side *AB*?

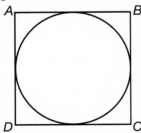

a. 10 inches
b. 20 inches
c. 100 inches
d. 400 inches

904. Gilda is making a quilt. She wants a quilt that is 30 square feet. She has collected fabric squares that are six inches by six inches. How many squares will she need?

a. 60

b. 90

c. 100

d. 120

905. What is the area of the following rectangle?

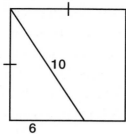

a. 30

b. 60

c. 64

d. 150

906. What is the value of X in the following figure?

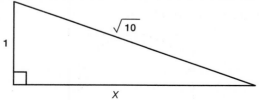

a. 3

b. 4

c. 5

d. 9

907. What is the area of the shaded figure inside the rectangle?

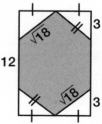

a. 18

b. 54

c. 60

d. 72

908. Ahmed has a canvas frame that is 25 inches long and 18 inches wide. When he centers it on top of a rectangular piece of canvas, the canvas is three inches longer on all four sides. What is the area of the canvas?

a. 450 square inches

b. 744 square inches

c. 588 square inches

d. 872 square inches

909. Kristen has a four-foot leash for her dog. She tied the leash to a stake in the ground so her dog would have some room to wander while she was working in the yard. What is the approximate area that her dog can roam while tied to the stake?

a. 13 square feet

b. 16 square feet

c. 25 square feet

d. 50 square feet

910. Zelda's book fell on the floor and landed with its spine up in an isosceles triangle. The height of the book triangle is $7\frac{1}{2}$ inches, and the base is $5\frac{1}{2}$ inches. What is the area of the triangle formed by the tossed book?

a. 41.25 square inches

b. 20.6 square inches

c. 32.5 square inches

d. 36.75 square inches

911. How many 12-inch square tiles are needed to tile the floor in a room that is 10 feet by 15 feet?

a. 150

b. 300

c. 144

d. 1,800

912. A city building needs a new emblem for the front lobby. The emblem is circular with an area of 452.16 square inches. Assuming π is 3.14, what is the radius of the emblem?

a. 6 inches

b. 12 inches

c. 144 inches

d. 288 inches

▶ **Set 58** (Answers begin on page 212.)

913. A two-story house is 20 feet high. The sides of the house are 28 feet long; the front and back are each 16 feet long. A gallon of paint will cover 440 square feet. How many gallons are needed to paint the whole house?
   a. 3 gallons
   b. 4 gallons
   c. 5 gallons
   d. 6 gallons

914. If the perimeter of a rectangle is 40 centimeters and the shorter sides are 4 centimeters, what is the length of the longer sides?
   a. 12 centimeters
   b. 10 centimeters
   c. 18 centimeters
   d. 16 centimeters

915. A square television has an area of 676 square inches. How long is each side?
   a. 26 inches
   b. 25 inches
   c. 24 inches
   d. 23 inches

916. The kitchen in Dexter's old house contains a circular stone inset with a radius of three feet in the middle of the floor. The room is square; each wall is 11.5 feet long. Dexter wants to re-tile the kitchen, all except for the stone inset. What is the approximate square footage of the area Dexter needs to buy tiles for?
   a. 76 square feet
   b. 28 square feet
   c. 132 square feet
   d. 104 square feet

917. The new hotel in the city is replacing an old lobby window. The window opening is 18 feet by 32 feet. What is the size of the piece of glass needed to fill the hole completely?
   a. 100 square feet
   b. 50 square feet
   c. 576 square feet
   d. 1,152 square feet

918. Kari is running for student council. The rules restrict the candidates to four two-foot-by-three-foot posters. Kari has dozens of four-inch-by-six-inch pictures that she would like to cover the posters with. What is the maximum number of pictures she will be able to use on the four posters?
   a. 144
   b. 130
   c. 125
   d. 111

919. Eli has a picture that is 25 inches by 19 inches and a frame that is 30 inches by 22 inches. He will make up the difference with a mat that will fit between the picture and the frame. What is the area of the mat he will need?
   a. 660 square inches
   b. 475 square inches
   c. 250 square inches
   d. 185 square inches

920. The area of a triangle is 54 square centimeters. The base is 18 centimeters. What is the height of the triangle?
   a. 6 centimeters
   b. 3 centimeters
   c. 9 centimeters
   d. 12 centimeters

921. Larry needs 45 square yards of fabric for curtains on two windows. If the fabric is 54 inches wide, how many yards should he buy?
  a. 10 yards
  b. 15 yards
  c. 20 yards
  d. 30 yards

922. The ceiling in a rectangular room is eight feet high. The room is 15 feet by 12 feet. Bernie wants to paint the long walls Moss and the short walls Daffodil. The paint can be purchased in gallon cans only, and each gallon will cover 200 square feet. How many cans of each color will Bernie need?
  a. 2 cans of Moss and 1 can of Daffodil
  b. 1 can of Moss and 2 cans of Daffodil
  c. 2 cans of Moss and 2 cans of Daffodil
  d. 1 can of Moss and 1 can of Daffodil

923. A doctor's office waiting room needs to be recarpeted. The carpeting costs $23.50 per square foot. How much would it cost to replace the carpet if the room's dimensions are 14 feet by 20 feet?
  a. $6,058
  b. $6,085
  c. $6,850
  d. $6,580

924. A rectangular school hallway is to be tiled with six-inch-square tiles. The hallway is 72 feet long and 10 feet wide. Lockers along both walls narrow the hallway by one foot on each side. How many tiles are needed to cover the hallway floor?
  a. 16
  b. 20
  c. 2,304
  d. 2,880

925. Mike is looking for a place on his wall to hang a portrait of his children. The framed portrait is 24 inches long, and its area is 432 square inches. How wide of a space on the wall does Mike need to have in order for the portrait to fit?
  a. 12 inches
  b. 14 inches
  c. 18 inches
  d. 20 inches

926. A waiter folds a napkin in half to form a five-inch square. What is the area, in square inches, of the unfolded napkin?
  a. 25 square inches
  b. 30 square inches
  c. 50 square inches
  d. 60 square inches

927. Shoshanna has a lamp placed in the center of her yard. The lamp shines a radius of 10 feet on her yard, which is 20 feet on each side. How much of the yard, in square feet, is NOT lit by the lamp?
  a. 400π square feet
  b. 100π square feet
  c. $40 - 10\pi$ square feet
  d. $400 - 100\pi$ square feet

928. In the diagram, a half circle is laid adjacent to a triangle. What is the total area of the shape, if the radius of the half circle is three and the height of the triangle is four?

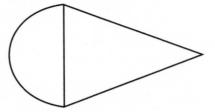

  a. $6(\pi + 4)$
  b. $6\pi + 12$
  c. $6\pi + 24$
  d. $\frac{9\pi}{2} + 12$

► **Set 59** (Answers begin on page 213.)

929. Juana is building a mall with 20 stores, each 20 by 35 feet. The mall has one hallway with dimensions of 100 by 20 feet. What is the mall's square footage?
  a. 18,000
  b. 16,000
  c. 6,000
  d. 2,000

930. Tamika is restoring an antique storage chest that is in the shape of a rectangular box. She is painting only the outside of the trunk. The trunk is four feet long, 18 inches wide, and two feet tall. There is a brass ornament on the outside of the trunk that takes up an area of one square foot that will not get painted. How much paint does she need, in square feet, to cover the outside of the trunk?
  a. 33 square feet
  b. 34 square feet
  c. 143 square feet
  d. 231 square feet

931. If the radius of a circle is seven centimeters, what is the area of the circle?
  a. 14π
  b. 21π
  c. 49π
  d. 7π

932. The circumference of a circle is 131.88. Assuming π is 3.14, what is the diameter of the circle?
  a. 42 inches
  b. 44 inches
  c. 46 inches
  d. 48 inches

933. Emily is wallpapering a circular room with a 10-foot radius and a height of eight feet. Ignoring the doors and windows, how much wallpaper will she need?
  a. 252π square feet
  b. 70π square feet
  c. 100π square feet
  d. 160π square feet

934. Mary wants to know the distance around the playground that she runs around three times a week. The width of the field is 75 meters, and the length is 150 meters.
  a. 11,250 square meters
  b. 450 square meters
  c. 11,250 meters
  d. 450 meters

935. The base of a triangle is twice the height of the triangle. If the area is 16 square inches, what is the height?
  a. 4 inches
  b. 8 inches
  c. 12 inches
  d. 16 inches

936. The area of a rectangular poster is 330 square centimeters. If the length is 22 centimeters, what is the width?

   a. 14 centimeters

   b. 15 centimeters

   c. 16 centimeters

   d. 17 centimeters

937. What is the area of the following isosceles triangle?

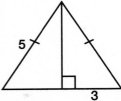

   a. 12 square units

   b. 15 square units

   c. 6 square units

   d. 24 square units

938. What is the area of the following parallelogram?

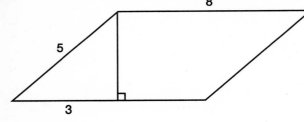

   a. 15

   b. 24

   c. 32

   d. 40

939. What is the area of the following figure?

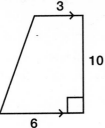

   a. 30 square units

   b. 45 square units

   c. 60 square units

   d. 90 square units

940. What is the measure of the diagonal of the following rectangle?

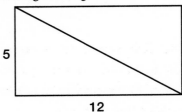

   a. 8.5

   b. 12

   c. 13

   d. 17

941. Find the area of the shaded portion in the following figure.

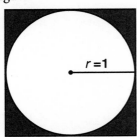

   a. π

   b. $\pi - 1$

   c. $2 - \pi$

   d. $4 - \pi$

942. What is the area of the following figure?

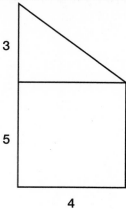

a. 22
b. 26
c. 30
d. 32

943. What is the difference in area between a square with a base of four inches; and a parallelogram with a base of four inches, a slant of 6 inches, and a height of four inches?

a. 0 square inches
b. 2 square inches
c. 4 square inches
d. 8 square inches

944. The radius of a given circle is six meters. What is the area of the circle?

a. 3π
b. 6π
c. 12π
d. 36π

► **Set 60** (Answers begin on page 214.)

945. Horses are racing on a circular track with a perimeter of 360 feet. Two cameras are in the center of the track. Camera A is following Horse A, and Camera B is following Horse B. If the angle between the two cameras is 40°, how far apart are the two horses?
a. 80π feet
b. 40π feet
c. 40 feet
d. 90 feet

946. Lily is replacing all the plumbing in her house. The pipe she is using has a diameter of 2.5 inches. What is the circumference of the end of the pipe?
a. 7.85 inches
b. 15.7 inches
c. 5 inches
d. 6.28 inches

947. A horse is tied to a post with a 20-foot rope. What is the longest path that the horse can walk?
a. 20 feet
b. 40 feet
c. 20π feet
d. 40π feet

948. Rick is making a glass cover for the top of a round end table. The diameter of the table is 36 inches. What will be the approximate area of the glass cover for the tabletop?
a. 1,017.36 square inches
b. 4,071.50 square inches
c. 113.10 square inches
d. 1,296 square inches

949. Reina is using silk to decorate the lids of round tin boxes. She will cut the silk in circles and use fabric glue to attach them. The lids have a radius of $4\frac{1}{2}$ inches. What should be the approximate circumference of the fabric circles?
a. $14\frac{1}{4}$ inches
b. $18\frac{1}{2}$ inches
c. $24\frac{1}{2}$ inches
d. $28\frac{1}{4}$ inches

950. In order to fell a tree, lumberjacks drive a spike through the center. If the circumference of a tree is 43.96 feet, what is the minimum length needed to go completely through the tree, passing through the center?
a. 7 feet
b. 14 feet
c. 21 feet
d. 28 feet

951. Farmer Harris finds a crop circle in the north 40 acres, where he believes an alien spacecraft landed. A reporter from the *Daily News* measures the circle and finds that its radius is 25 feet. What is the circumference of the crop circle?
a. 78.5 feet
b. 157 feet
c. 208 feet
d. 357.5 feet

952. Jessie wants to buy a round dining table the same size as her circular rug. The circumference of the rug is 18.84 feet. What should be the diameter of the table she buys?
a. 6 feet
b. 18 feet
c. 3.14 feet
d. 7 feet

953. The area of a given circle is 49π. What is the radius?
- **a.** 7
- **b.** 9
- **c.** 11
- **d.** 13

954. Ana is making cookies. She rolls out the dough to a rectangle that is 18 inches by 12 inches. Her circular cookie cutter has a circumference of 9.42 inches. Assuming she reuses the dough scraps and all the cookies are the same weight, approximately how many cookies can Ana cut out of the dough?
- **a.** 31
- **b.** 14
- **c.** 12
- **d.** 10

955. Skyler's bike wheel has a diameter of 27 inches. He rides his bike until the wheel turns 100 times. How far did he ride?
- **a.** 1,500π inches
- **b.** 2,700π inches
- **c.** 3,200π inches
- **d.** 4,800π inches

956. Find the perimeter of an isosceles triangle that has a base of 8 centimeters and one side that measures 14 centimeters.
- **a.** 36 centimeters
- **b.** 36 square centimeters
- **c.** 22 centimeters
- **d.** 22 square centimeters

957. Find the perimeter of a regular octagon that has sides measuring 13 centimeters each.
- **a.** 78 centimeters
- **b.** 91 centimeters
- **c.** 104 centimeters
- **d.** 117 centimeters

958. Bridget wants to hang a garland of silk flowers all around the ceiling of a square room. Each side of the room is nine feet long; the garlands are available in 15-foot lengths only. How many garlands will she need to buy?
- **a.** 2
- **b.** 3
- **c.** 4
- **d.** 5

959. The perimeter of a rectangle is 38 centimeters. What are the dimensions of the rectangle?
- **a.** 10 centimeters × 6 centimeters
- **b.** 12 centimeters × 4 centimeters
- **c.** 12 centimeters × 2 centimeters
- **d.** 10 centimeters × 9 centimeters

960. Michelle painted an eight-foot-by-10-foot canvas floor cloth with a blue border eight inches wide. What is the perimeter of the unpainted section?
- **a.** $30\frac{2}{3}$ feet
- **b.** $15\frac{1}{3}$ feet
- **c.** $35\frac{1}{3}$ feet
- **d.** $20\frac{2}{3}$ feet

▶ **Set 61** (Answers begin on page 215.)

961. Emily would like to make a new pillowcase for her pillow. The length of the pillow is 2.5 feet and the width is 1.5 feet. What is the area of the pillow?
   a. 4 square feet
   b. 3.75 square feet
   c. 3.25 square feet
   d. 3 square feet

962. Minako has a fish pond shaped like an equilateral triangle, that is 20 feet on a side. She wants to fill the pond, which is 10 feet deep, with water. How much water will she need?
   a. $100\sqrt{5}$ cubic feet
   b. $100\sqrt{3}$ cubic feet
   c. 1,000 cubic feet
   d. $1,000\sqrt{3}$ cubic feet

963. Juan takes a compass reading and finds that he is 32° north of east. If he was facing the exact opposite direction, what would his compass reading be?
   a. 32° north of west
   b. 58° west of south
   c. 58° north of east
   d. 32° south of east

964. Renee is putting a stairway in her house. If she wants the stairway to make an angle of 20° to the ceiling, what obtuse angle should the stairway make to the floor?
   a. 30°
   b. 40°
   c. 110°
   d. 160°

965. When building a house, the builder should make certain that the walls meet at a 90° angle. This is called
   a. a straight angle.
   b. an acute angle.
   c. an obtuse angle.
   d. a right angle.

966. Zelda cuts a pizza in half in a straight line. She then cuts a line from the center to the edge, creating a 35° angle. What is the supplement of that angle?
   a. 55°
   b. 145°
   c. 35°
   d. 70°

967. In a pinball machine, a ball bounces off a bumper at an angle of 72° and then bounces off another bumper that is parallel to the first. If the angles that the ball comes to and leaves the first bumper from are equal, what is the angle that the ball bounces off the second bumper?

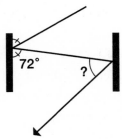

   a. 36°
   b. 48°
   c. 72°
   d. 90°

968. To wash the windows on the second floor of his house, Alan leaned a ladder against the house, creating a 54° acute angle with the ground. What is the measure of the angle the ladder made with the house?

  a. 36°
  b. 126°
  c. 81°
  d. 306°

969. A wheelchair ramp must have a run (horizontal distance) of one foot for every inch that the ramp goes up. How long (diagonal distance) must a ramp be built in order to cover the height of an eight-inch step?

  a. $8\sqrt{145}$ inches
  b. $8\sqrt{2}$ inches
  c. $4\sqrt{13}$ inches
  d. 9,280 inches

970. Which of the following statements best describes the vertex of an angle?

  a. the longest side of an angle
  b. the imaginary third side of an angle
  c. the point where the two rays of an angle meet
  d. the formula for finding the degrees in an angle

971. A chicken is standing at the side of a straight section of road. On the same side, a cow is standing 60 feet down the road; on the other side of the road and directly across from the chicken, a bowl of corn sits. The corn is 80 feet from the cow. About how far will the chicken have to go to cross the road to eat the corn? (Note: A calculator is needed for this problem.)

  a. 45 feet
  b. 49 feet
  c. 53 feet
  d. 61 feet

972. Two ships leave from a port. Ship A sails west for 300 miles, and Ship B sails north for 400 miles. How far apart are the ships after their trips?

  a. 300 miles
  b. 400 miles
  c. 500 miles
  d. 900 miles

973. What is the measure of the arc of a circle intercepted by a central angle that measures 64°?

  a. 32°
  b. 64°
  c. 128°
  d. 360°

974. Rachel has a round coffee table that has six equal triangle-shaped inlays. The smallest angle of each triangle meets in the center of the table; each triangle touches the edge of the triangle on either side. Three intersecting lines, therefore, create the six identical triangles. What is the angle of each of the points in the center of the table?

    a. 120°

    b. 90°

    c. 75°

    d. 60°

975. The most ergonomically correct angle between the keyboard and the screen of a laptop computer is 100°. This is called

    a. an acute angle.

    b. a complimentary angle.

    c. an obtuse angle.

    d. a right angle.

976. When Ivan swam across a river, the current carried him 20 feet downstream. If the total distance that he swam was 40 feet, how wide is the river to the nearest foot?

    a. 30 feet

    b. 32 feet

    c. 35 feet

    d. 50 feet

▶ **Set 62** (Answers begin on page 216.)

977. Lara is standing next to a tree with a shadow of 40 feet. If Lara is 5.5 feet tall with a shadow of 10 feet, what is the height of the tree?
- **a.** 10 feet
- **b.** 11 feet
- **c.** 20 feet
- **d.** 22 feet

978. Stefan is leaning a ladder against his house so that the top of the ladder makes a 75° angle with the house wall. What is the measure of the acute angle where the ladder meets the level ground?
- **a.** 15°
- **b.** 25°
- **c.** 45°
- **d.** 165°

979. Greta is building a path in her yard. She wants the path to bisect her garden on the side that is 36 feet long. How far from the end of Greta's garden will the path cross it?
- **a.** 12 feet
- **b.** 18 feet
- **c.** 20 feet
- **d.** 72 feet

980. Mario the bike messenger picks up a package at Print Quick, delivers it to Bob's office 22 blocks away, waits for Bob to approve the package, and then hops on his bike and returns the package to Print Quick. If one block is 90 yards, how many yards will Mario bike in total?
- **a.** 44 yards
- **b.** 180 yards
- **c.** 1,980 yards
- **d.** 3,960 yards

981. Dennis is building a brick wall that will measure 10 feet by 16 feet. If the bricks he is using are three inches wide and five inches long, how many bricks will it take to build the wall?
- **a.** 128
- **b.** 160
- **c.** 1,536
- **d.** 23,040

982. A power line stretches down a 400-foot country road. A pole is to be put at each end of the road with one in the midpoint of the wire. How far apart is the center pole from the left-most pole?
- **a.** 100 feet
- **b.** 200 feet
- **c.** 300 feet
- **d.** 400 feet

983. Justin walks to school in a straight line; the distance is seven blocks. The first block is 97 feet long; the second and third blocks are 90 feet long; the fourth, fifth, and sixth blocks are 110 feet long. The seventh block is congruent to the second block. How far does Justin walk?

    a. 770 feet

    b. 717 feet

    c. 704 feet

    d. 607 feet

984. Lines *A* and *B* are parallel to each other. Lines *C* and *D* are also parallel to each other, and in addition, both perpendicular to lines *A* and *B*. Suppose line *E* is a transversal to line *A*. What other line(s) must line *E* cross?

    a. Line *B*

    b. Line *B* and *C*

    c. Line *B* and *D*

    d. Line *C* and *D*

985. Betty Lou, Norma Jean, Margery, and Dave want to share a brownie evenly and only cut the brownie twice. Margery cuts one line down the middle and hands the knife to Dave. What kind of line should Dave cut through the middle of the first line?

    a. a parallel line

    b. a transversal line

    c. a perpendicular line

    d. a skewed line

986. Oddly, the town of Abra lies half in one state and half in another. To make it even more confusing, the straight line that divides the states runs along Sandusky Street. What is the relationship of the houses at 612 Sandusky, 720 Sandusky, and 814 Sandusky?

    a. The houses are congruent.

    b. The houses are collinear.

    c. The houses are segmented.

    d. The houses are equilateral.

987. What is the sum of the measures of the exterior angles of a regular pentagon?

    a. 72°

    b. 108°

    c. 360°

    d. 540°

988. The city requires all pools to be in fenced-in yards. Rick's yard is 60 feet by 125 feet. How much fencing does he need to fence in the yard for the pool, excluding a four-foot gate at the sidewalk?

    a. 181 feet

    b. 366 feet

    c. 370 feet

    d. 7,496 feet

989. If two sides of a triangle measure five and seven, between what two numbers must the length of the third side be?

    a. 2 and 5

    b. 2 and 12

    c. 5 and 7

    d. The third side can be any length.

990. A container filled with water is 10 by 10 by 15 inches. Portia fills her glass with 60 cubic inches of water from the container. How much water is left in the container?
- **a.** 1,500 cubic inches
- **b.** 1,440 cubic inches
- **c.** 1,000 cubic inches
- **d.** 60 cubic inches

991. Find the volume of a box with a length of 8.5 inches, a width of 5.5 inches, and a height of 10 inches.
- **a.** 457.6 cubic inches
- **b.** 476.5 cubic inches
- **c.** 467.5 cubic inches
- **d.** 475.6 cubic inches

992. A water tank is in the shape of a right circular cylinder. The base has a radius of six feet. If the height of the tank is 20 feet, what is the approximate volume of the water tank?
- **a.** 377 cubic feet
- **b.** 720 cubic feet
- **c.** 754 cubic feet
- **d.** 2,261 cubic feet

▶ **Set 63** (Answers begin on page 216.)

993. A builder has 27 cubic feet of concrete to pave a sidewalk whose length is six times its width. The concrete must be poured six inches deep. How long is the sidewalk?

a. 9 feet
b. 12 feet
c. 15 feet
d. 18 feet

994. Anne has two containers for water: A rectangular plastic box with a base of 16 square inches, and a cylindrical container with a radius of two inches and a height of 11 inches. The rectangular box is filled with water nine inches from the bottom. If Anne pours all this water into the cylindrical container, which of the following will be true?

a. The cylinder will overflow.
b. The cylinder will be exactly full.
c. The cylinder will be filled to an approximate level of 10 inches.
d. The cylinder will be filled to an approximate level of 8 inches.

995. A cereal box is filled with Snappie cereal at the Snappie production facility. When the box arrives at the supermarket, the cereal has settled and takes up a volume of 144 cubic inches. There is an empty space in the box of 32 cubic inches. If the base of the box is two by eight inches, how tall is the box?

a. 5 inches
b. 11 inches
c. 12 inches
d. 15 inches

996. Sunjeev has a briefcase with dimensions of two by $1\frac{1}{2}$ by $\frac{1}{2}$ feet. He wishes to place several notebooks, each eight by nine by one inch, into the briefcase. What is the largest number of notebooks the briefcase will hold?

a. 42
b. 36
c. 20
d. 15

997. One cubic foot of copper is to be made into wire with a $\frac{1}{2}$ inch radius. If all of the copper is used to produce the wire, how long will the wire be, in inches?

a. 2,000 inches
b. $\frac{4,000}{\pi}$ inches
c. $\frac{6,912}{\pi}$ inches
d. $\frac{48,000}{\pi}$ inches

998. Daoming wants to know the height of a telephone pole. He measures his shadow, which is three feet long, and the pole's shadow, which is 10 feet long. Daoming's height is six feet. How tall is the pole?

a. 40 feet
b. 30 feet
c. 20 feet
d. 10 feet

999. Which type of transformation can change the size of an object?
- **a.** reflection
- **b.** dilation
- **c.** translation
- **d.** rotation

1000. Which three points are both collinear and coplanar?

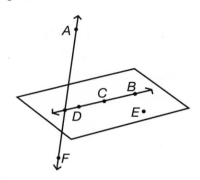

- **a.** *A, B, C*
- **b.** *A, C, D*
- **c.** *B, C, D*
- **d.** *D, E, A*

1001. A pole that casts a 15-foot-long shadow stands near an eight-foot-high stop sign. If the shadow cast by the sign is three feet long, how high is the pole?
- **a.** 40 feet
- **b.** 30 feet
- **c.** 24 feet
- **d.** 45 feet

▶ Answers

▶ Section 1— Miscellaneous Math

Set 1 (Page 2)

1. **c.** The correct answer to this basic addition problem is 14.

2. **b.** The correct answer to this basic subtraction problem is 6.

3. **a.** The correct answer to this basic division problem is 3.

4. **d.** The correct answer to this basic multiplication problem is 16.

5. **b.** The correct answer to this basic subtraction problem is 44.

6. **a.** The correct answer to this basic addition problem is 100.

7. **a.** The correct answer to this basic addition problem is 49.

8. **c.** The correct answer to this basic addition problem is 100.

9. **a.** The correct answer to this basic subtraction problem is 43.

10. **b.** The correct answer to this basic subtraction problem is 82,202.

11. **b.** In this case (but not in all cases, as you will see in some of the following problems), perform the operations in the order presented, from left to right: $72 + 98 = 170$; then, $170 - 17 = 153$.

12. d. Again, do this problem in the order presented. First subtract, then add: 353 is the correct answer.

13. c. First add, then subtract. In multistep problems, be careful not to rush just because the operations are simple. The correct answer is 560.

14. b. First add, then subtract. The correct answer is 6,680. When doing problems involving several digits, it's easy to make mistakes. Be sure to read carefully, no matter how simple the problem seems.

15. a. First subtract, then add. The correct answer is 5,507.

16. c. The bars on either side of 7 indicate "the absolute value of 7." The *absolute value* of a number is the distance that number is away from zero on a number line. The absolute value of 7 is 7.

Set 2 (Page 4)

17. b. Perform the operations within the parentheses first: $15 + 32 = 47$; $56 - 39 = 17$. Then, multiply those sums as indicated by the adjoined parentheses: $47 \times 17 = 799$.

18. a. The correct answer to this division problem is 9.

19. d. The correct answer to this multiplication problem is 25,200.

20. c. The correct answer to this multiplication problem is 15,050.

21. a. The correct answer to this multiplication problem is 7,290.

22. d. $56,515 \div 4 = 14128.75$, or, rounded to the nearest whole number, 14,129.

23. d. The bars on either side of –9 indicate "the absolute value of –9." The *absolute value* of a number is the distance that number is away from zero on a number line. The absolute value of –9 is 9 because it is nine units away from zero. Because absolute value is a distance, it is always positive.

24. b. Perform the operation within the parentheses first: $6 + 3 = 9$; then, multiply the sum by 5. The correct answer is 45.

25. a. The correct answer is 15. An incorrect answer may represent an error in place value.

26. c. The correct answer is 12,407. If you chose choice **a**, you disregarded the zero in 62,035.

27. c. The answer when the two numbers are divided is approximately 100.03, which is closest to 100. The question can also be evaluated using compatible numbers, such as $6,300 \div 63$; and the solution here is also 100.

28. a. The correct answer is 205,335.

29. a. Perform the operations in parentheses first: $84 - 5 = 79$ from the first set and $3 \times 54 = 162$ from the second set. The expression is now $12(79) - 162$. Multiply 12 times 79 to get 948, and then subtract: $948 - 162 = 786$.

30. b. The correct answer to this multiplication problem is 250,608.

31. d. Perform the operation within the parentheses first: $12 \times 6 = 72$; then, add the product by 18. The correct answer is 90.

32. a. First multiply, then add. The correct answer is 1,467.

Set 3 (Page 6)

33. **c.** First multiply; then subtract. The correct answer is 65,011.

34. **c.** Divide first; then add. The answer is 162.

35. **a.** In spite of the sequence presented, in this problem you must divide first: $204 \div 2 = 102$. Now subtract 102 from 604: $604 - 102 = 502$. The correct answer is 502. When an expression involves more than one operation, you need to follow the correct order of operations: 1) Operations in **P**arentheses; 2) **E**xponents; 3) **M**ultiplication and **D**ivision in order from left to right; 4) **A**ddition and **S**ubtraction in order from left to right. A good device for remembering this order is saying: **P**lease **E**xcuse **M**y **D**ear **A**unt **S**ally.

36. **d.** Do the operation in parentheses first: $202 \div 2 = 101$; then, subtract the answer from 604: $604 - 101 = 503$.

37. **b.** Multiply twice. The correct answer is 14,560.

38. **d.** Multiply within the parentheses first: $8 \times 3 = 24$; then, multiply the product by 17. The correct answer is 408.

39. **d.** Use the place value of each of the nonzero numbers. The four is in the ten thousands place, so it is equal to 40,000, the five is in the hundreds place, so it is equal to 500, and the three is in the ones place, so it is equal to 3; $40,000 + 500 + 3 = 40,503$.

40. **c.** The number 112 is divisible by both seven and eight because each can divide into 112 without a remainder; $112 \div 7 = 16$ and $112 \div 8 = 14$. Choice **a** is only divisible by seven, choice **b** is not divisible by either, and choice **d** is only divisible by eight.

41. **b.** A prime number is a whole number whose only factors are one and itself. Two, three, and seven are all prime numbers. The prime factors of a number are the prime numbers that multiply to equal that number: $2 \times 3 \times 7 = 42$.

42. **a.** Forty-six goes into 184 four times. The other choices cannot be divided evenly into 184.

43. **c.** Subtract the pounds first, and then the ounces.

44. **d.** You must "borrow" 60 minutes from the three hours in order to be able to subtract.

45. **b.** Add the four numbers together for a sum of 244. Then, divide by 4 to get a quotient of 61.

46. **c.** Divide the larger number by the smaller number for a quotient of 4,978.

47. **a.** Add the feet first, and then the inches: $3 + 5 = 8$; $6 + 8 = 14$. Then, convert 12 inches to 1 foot with 2 inches leftover. Add to 8 feet. The correct answer is 9 feet 2 inches.

48. **c.** Add the hours first, and then the minutes: 1 hour + 3 hours = 4 hours. 20 minutes + 30 minutes = 50 minutes. Combine: 4 hours 50 minutes.

Set 4 (Page 8)

49. **a.** The number 157 is rounded to 200, and 817 is rounded to 800; $200 \times 800 = 160,000$.

50. **b.** When *adding* negative numbers, follow this rule: If both numbers have **different signs**, subtract the smaller from the larger. The sum takes the sign of the larger number. In this equation the correct answer is -13.

51. c. When multiplying integers, begin by multiplying the numbers: $4 \times 9 = 36$. If there is an *odd* number of negative signs, the answer is negative. If there is an *even* number of negative signs, the answer is positive. In this equation, there are two negative signs; therefore, the correct answer is 36.

52. a. When adding negative numbers, simply add the numbers and carry the negative sign into the answer. The correct answer to this equation is -12.

53. c. When adding a negative number to a positive number, subtract the smaller from the larger and carry the sign of the larger number into the answer. The correct answer to this equation is -8.

54. a. When adding two numbers that are the same, but have opposite signs, you will always get zero as your answer.

55. c. The meaning of 6^3 is *6 to the power of 3*, or 6 used as a factor 3 times: $6 \times 6 \times 6 = 6^3$.

56. d. In this problem, 2 is the base and 5 is the exponent. *Two raised to the power of 5 means* to use 2 as a factor five times: $2 \times 2 \times 2 \times 2 \times 2 = 32$.

57. d. A composite number is a whole number greater than 1 that has other factors besides 1 and itself; in other words, it is not prime. Each of the answer choices are prime numbers except 91, which has factors of 1, 7, 13, and 91.

58. d. When squaring a number, simply multiply it by itself: $19 \times 19 = 361$.

59. b. To solve this division problem, subtract the exponents only: $5 - 2 = 3$, so the answer is 10^3.

60. c. -4^2 means the opposite of the result of 4 raised to the second power. Think of it as -1×4^2. As in the order of operations, evaluate the exponent first, and then multiply: $-1 \times 16 = -16$.

61. d. -5^3 means the opposite of the result of 5 raised to the third power. Think of it as -1×5^3. As in the order of operations, evaluate the exponent first, and then multiply: $-1 \times (5 \times 5 \times 5) = -1 \times 125 = -125$.

62. a. Since there are parentheses, the base in this question is -11; $(-11)^2$ means -11×-11, which is equal to 121. Because there is an even number of negatives, the answer is positive.

63. a. To find the square root of a number, ask the question, "What number times itself equals 64?" The answer is 8.

64. a. Square roots can be multiplied and divided, but they cannot be added or subtracted.

Set 5 (Page 10)

65. c. The square root of 64 is 8.

66. d. The symbol "!" means *factorial*, which is the product of that number and **every** positive whole number less than it: $5! = 5 \times 4 \times 3 \times 2 \times 1 = 120$.

67. b. Divide the total number of hot dogs (400) by the amount in each package (8), to get the number of packages: $400 \div 8 = 50$.

68. d. Begin by adding: 12 ounces + 10 ounces = 22 ounces. Since this unit of measurement is not an answer choice, convert to pounds. There are 16 ounces in 1 pound, so 22 ounces is equal to 1 pound 6 ounces.

69. b. Change the hours to minutes: 1 hour 40 minutes = 100 minutes; 1 hour 50 minutes = 110 minutes. Now add: 100 minutes + 110 minutes = 210 minutes. Now change back to hours and minutes: 210 minutes ÷ 60 minutes = 3.5 hours.

70. d. Multiply the number of members (35) by the average number of bars per person (6): 35 × 6 = 210.

71. a. To answer this question, subtract each amount of purchase from the $58 she started with; $58 − $18 = $40 and then $40 − $6 = $34. She has $34 left.

72. b. Add the amount of miles for each day for a total of 696 miles; 696 rounded to the nearest ten or nearest hundred is 700.

73. c. To find their total score, add their individual scores together: 189 + 120 + 120 + 95 = 524. Don't forget to add 120 twice for both Charles *and* Max.

74. c. You must subtract the reading at the beginning of the week from the reading at the end of the week: 21,053 − 20,907 = 146.

75. c. The values added together total $618. If you chose choice **a**, you forgot that the value of the handbag ($150) must also be included in the total.

76. b. This is a basic addition problem: 108 pounds + 27 pounds = 135 pounds.

77. c. The total value is $5,525. It is important to remember to include all three telephone sets ($375 total), both computers ($2,600 total), and both monitors ($1,900 total) in the total value.

78. b. Add the value of the three sweaters (3 × 68 = 204), the computer game after the rebate (75 − 10 = 65) and one bracelet (43); 204 + 65 + 43 = $312.

79. a. This is a two-step subtraction problem. First you must find out how many miles the truck has traveled since its last maintenance. To do this, subtract: 22,003 − 12,398 = 9,605. Now subtract 9,605 from 13,000 to find out how many more miles the truck can travel before it must have another maintenance: 13,000 − 9,605 = 3,395.

80. d. This is a problem involving multiplication. The easiest way to solve this problem is to temporarily take away the five zeros, then multiply: 365 × 12 = 4,380. Now add back the five zeros for a total of 438,000,000. (If you selected choice **a**, you mistakenly divided when you should have multiplied.)

Set 6 (Page 13)

81. c. First ask how many inches are in one foot; the answer is 12 inches. Now multiply: 12 × 4 = 48 inches.

82. c. This is a division problem: 46 ÷ 2 = 23. Rajeeve is 23 years old.

83. a. This is a two-step division problem: 2,052 miles ÷ 6 days = 342 miles per day; 342 miles per day ÷ 2 stops = 171 miles between stops.

84. b. Shoshanna's three best (that is, lowest) times are 54, 54, and 57, which add up to 165. Now divide to find the average of these times: 165 ÷ 3 = 55. If you got the wrong answer, you may have added all of Shoshanna's times, rather than just her best three. Even when the problem seems simple and you're in a hurry, be sure to read carefully.

85. **d.** To find the average, divide the total number of miles, 3,450, by 6 days: 3,450 miles ÷ 6 days = 575 miles per day.

86. **c.** First, find the total hours worked by all six nurses: 8 + 10 + 9 + 8 + 7 + 12 = 54. Then find the average by dividing the total hours by the number of nurses: 54 ÷ 6 = 9.

87. **d.** Take the total number of miles and find the average by dividing: 448 miles ÷ 16 gallons = 28 miles per gallon.

88. **d.** 827,036 bytes free + 542,159 bytes freed when the document was deleted = 1,369,195 bytes; 1,369,195 bytes − 489,986 bytes put into the new file = 879,209 bytes left free.

89. **c.** The solution to the problem entails several operations: First, multiply $80 per month by 7 months = $560. Next, multiply $20 per month by the remaining 5 months = $100. Now find the average for the entire year. Add the two amounts: $560 + $100 = $660. Now divide: $660 ÷ 12 months in a year = $55.

90. **c.** This is a two-step problem. First, add the three numbers: 22 + 25 + 19 = 66. Now divide the sum by 3 to find the average: 66 ÷ 3 = 22.

91. **c.** First, convert feet to inches: 3 feet = 3 × 12 inches = 36 inches. Now add 4 inches: 36 inches + 4 inches = 40 inches. Then do the final operation: 40 inches ÷ 5 = 8 inches.

92. **c.** Multiply 16 by 5 to find out how many gallons all five sprinklers will release in one minute. Then multiply the result (80 gallons per minute) by the number of minutes (10) to get the entire amount released: 80 × 10 = 800 gallons.

93. **b.** The median value is the middle value in the list when the numbers are arranged in ascending or descending order. In ascending order

this list becomes 8, 8, 9, 10, 12, 13, 17. The middle number is 10.

94. **d.** It will take one worker about twice as long to complete the task, so you must multiply the original hours and minutes times two: 2 hours 40 minutes × 2 = 4 hours 80 minutes, which is equal to 5 hours 20 minutes.

95. **b.** Two candy bars require two quarters; one package of peanuts requires three quarters; one can of cola requires two quarters—for a total of seven quarters.

96. **b.** If it is 11:46 A.M., in 14 minutes it will be noon. In 15 minutes, then, it will be 12:01 P.M.

Set 7 (Page 15)

97. **a.** To arrive at the answer quickly, begin by rounding off the numbers, and you will see that choice **a** is less than 600 pounds, whereas choices **b**, **c**, and **d** are all more than 600 pounds.

98. **b.** Take the number of classroom hours and divide by the number of courses: 48 ÷ 3 = 16 hours per course. Now multiply the number of hours taught for one course by the pay per hour: 16 × $35 = $560.

99. **b.** Carmella worked 15 hours per week for 8 weeks: 15 × 8 = 120. In addition, she worked 15 hours for Mariah for one week, so 120 + 15 = 135.

100. **c.** Divide the amount of cod by the number of crates: 400 ÷ 20 = 20.

101. **c.** Take apart the statement and translate each part. The word *Twice* tells you to multiply the quantity by two. In the second part of the statement, the word *sum* is a key word for addition. So *the sum of six and four* is translated as 6 + 4. The whole statement becomes 2(6 + 4).

102. **b.** The average is the sum divided by the number of observations: $(11 + 4 + 0 + 5 + 4 + 6) \div 6 = 5$.

103. **a.** The labor fee ($25) plus the deposit ($65) plus the basic service ($15) equals $105. The difference between the total bill, $112.50, and $105 is $7.50, the cost of the news channels.

104. **c.** Three tons = 6,000 pounds; 6,000 pounds × 16 ounces per pound = 96,000 ounces that can be carried by the truck. The total weight of each daily ration is: 12 ounces + 18 ounces + 18 ounces = 48 ounces per soldier per day; 96,000 ÷ 48 = 2,000; 2,000 ÷ 10 days = 200 soldiers supplied.

105. **c.** Between 8:14 and 9:00, 46 minutes elapse, and between 9:00 and 9:12, 12 minutes elapse, so this becomes a simple addition problem: 46 + 12 = 58.

106. **d.** Begin by multiplying to find out how many forms one clerk can process in a day: 26 forms × 8 hours = 208 forms per day per clerk. Now find the number of clerks needed by dividing: 5,600 ÷ 208 = 26.9. Since you can't hire 0.9 of a clerk, you have to hire 27 clerks for the day.

107. **c.** The Counting Principle shows the number of ways an event can occur and tells you to take the total choices for each item and multiply them together: $3 \times 5 \times 2 = 30$.

108. **a.** This answer is in the correct order and is "translated" correctly: **Rachel had (=) 3 apples and ate (−) 1.**

109. **c.** This is a multiplication problem: 84,720 troops times 4 square yards of cloth equals 338,880 square yards of cloth required. (If you chose choice **a**, you mistakenly divided.)

110. **c.** From 5:16 P.M. to 7:16 A.M. the next day is 14 hours. An additional 42 minutes occurs between 7:16 A.M. and 7:58 A.M: (58 − 16 = 42).

111. **c.** The bars on either side of −14 indicate the absolute value of −14. The absolute value of a number is its distance away from zero on a number line and, in this case, is 14. Therefore, 14 + −5 = 9. Since the signs are different, subtract and keep the sign of the larger number.

112. **d.** According to the table, a pound in weight makes a difference of $0.64, or $0.04 per ounce over 4 pounds. Fruit that weighs 4 pounds 8 ounces will cost 8 × 0.04 or $0.32 more than fruit that costs 4 pounds. Therefore, the cost is $1.10 + 0.32 = $1.42.

Set 8 (Page 18)

113. **b.** To find the answer, begin by adding the cost of the two sale puppies: $15 + $15 = $30. Now subtract this amount from the total cost: $70 − $30 = $40 paid for the third puppy.

114. **c.** January is approximately $38,000; February is approximately 41,000; and April is approximately 26,000. These added together give a total of 105,000.

115. **d.** Because Linda pays with a check only if an item costs more than $30, the item Linda purchased with a check in this problem must have cost more than $30. If an item costs more than $30, then it must cost more than $25 (choice **d**), as well.

116. **d.** This series actually has two alternating sets of numbers. The first number is doubled, giving the third number. The second number has 4 subtracted from it, giving the fourth number. Therefore, the blank space will be 12 doubled, or 24.

117. d. Figure the amounts by setting up the following equations: First, S = $3 + $23 = $26. Now, B = ($1 × 5) + ($2 × 2) or $5 + $4 = $9. MR = $1 × 2 = $2; and D = $4 × 1 = $4. Now, add: $9 + $2 + $4 = $15. Now subtract: $26 − $15 = $11.

118. b. The total value of the supplies and instruments is found by adding the cost of each item: 1,200 + (2 × 350) + (3 × 55) + 235 + 125 + 75 = 1,200 + 700 + 165 + 235 +125 + 75. The total is $2,500.

119. c. Follow the order of operations and evaluate the exponent first. Since any nonzero number raised to the zero power is equal to 1, then $4^0 = 1$. Now multiply 5(1) = 5 to get the simplified value.

120. a. It is easiest to use trial and error to arrive at the solution to this problem. Begin with choice **a**: After the first hour, the number would be 20; after the second hour, 40; after the third hour, 80; after the fourth hour, 160; and after the fifth hour, 320. Fortunately, in this case, you need go no further. The other answer choices do not have the same outcome.

121. a. The unreduced ratio is 8,000:5,000,000; reduced, the ratio is 8:5,000. Now divide: 5,000 ÷ 8 = 625, for a ratio of 1:625.

122. c. This is a three-step problem involving multiplication, subtraction, and addition. First, find out how many fewer minutes George jogged this week than usual: 5 hours × 60 minutes = 300 minutes − 40 minutes missed = 260 minutes jogged. Now add back the number of minutes George was able to make up: 260 minutes + 20 + 13 minutes = 293 minutes. Now subtract again: 300 minutes − 293 = 7 minutes jogging time lost.

123. a. Since 4 and 6 are factors of 24, and 4 is a perfect square, $\sqrt{24} = \sqrt{4} \times \sqrt{6} = 2 \times \sqrt{6}$ or $2\sqrt{6}$.

124. d. Irrational numbers are nonrepeating, nonterminating decimals. They also cannot be written in fraction form without approximating. Choice **d** has a pattern, but the same exact pattern does not repeat each time, making it irrational. Choice **a** is a repeating decimal, choice **b** is a fraction, and choice **c** is a terminating decimal, so they are rational, not irrational.

125. d. First, write the problem in columns:

6 feet   5 inches
− 5 feet   11 inches

Now subtract, beginning with the right-most column. Since you cannot subtract 11 inches from 5 inches, you must borrow 1 foot from the 6 in the top left column, then convert it to inches and add: 1 foot = 12 inches; 12 inches + 5 inches = 17 inches. The problem then becomes:

5 feet   17 inches
− 5 feet   11 inches
                6 inches

So the answer is choice **d**, 6 inches.

126. c. This is a problem of addition. You may simplify the terms: M = F + 10; then substitute: M = 16 + 10, or 26.

127. c. 76 ÷ 19 = 4; the other division operations will not end in whole numbers.

128. a. The mode of a set of numbers is the number that appears the most. Four appears in this set three times. Since it appears more often than any other number in the list, 4 is the mode.

Set 9 (Page 21)

129. **a.** Subtract the months first, then the years. Remember that it is best to write the problem in columns and subtract the right-most column (months) first, then the left-most column (years): 8 months − 7 months = 1 month; 2 years − 1 year = 1 year. So, Brook is 1 year 1 month older than Emilio.

130. **b.** First, simplify the problem: K = 5 × 17 = 85, so Kara made 85 cents; R = 7 × 14 = 98, so Rani made 98 cents, the higher amount of money; R − K = 98 − 85 = 13. Therefore, Rani made 13 cents more than Kara.

131. **a.** The train's speed can be found using the formula *Distance = rate × time*. From this we get the formula *rate = $\frac{distance}{time}$*, since we are looking for the speed. By substituting, *rate = $\frac{300}{6}$*, which simplifies to 50. The speed is 50 miles per hour.

132. **b.** First, divide to determine the number of 20-minute segments there are in 1 hour: 60 ÷ 20 = 3. Now multiply that number by the number of times Rita can circle the block: 3 × 5 = 15.

133. **a.** The symbol > means "greater than," and the symbol < means "less than." The only sentence that is correct is choice **a**: four feet is greater than three feet. The other choices are untrue.

134. **c.** A prime number is a number that has exactly two factors, 1 and itself. Each of the number choices has more than two factors except 11, whose only factors are 1 and 11.

135. **d.** If Salwa (S) is 10 years older than Roland (R), then S is equal to R + 10. Therefore, the equation is S = R + 10.

136. **d.** If four students are lining up, then there are four choices of students for the first spot in line, three choices for the second, two choices for the third, and one choice for the fourth spot. The Counting Principle tells you to take the possible choices and multiply them together: 4 × 3 × 2 × 1 = 24. This type of situation is also called a *permutation*, because the order matters.

137. **c.** First, simplify the problem: L = $5 + $4 + $6 = $15; R = $14. Lucinda spent the most by $1. Don't forget that only the popcorn was shared; the other items must be multiplied by two.

138. **c.** First, convert tons to pounds; 1 ton = 2,000 pounds; 36 tons (per year) = 72,000 pounds (per year); 1 year = 12 months, so the average number of pounds of mosquitos the colony of bats can consume in a month is: 72,000 ÷ 12, or 6,000 pounds.

139. **a.** In this question, 10 is the base and 4 is the exponent; 10^4 means 10 is used as a factor four times, or 10 × 10 × 10 × 10.

140. **c.** This is a three-step problem involving multiplication and division. First, change yards to feet: 240 yards × 3 feet in a yard = 720 feet. Now find the number of square feet in the parcel: 121 feet × 740 feet = 87,120 square feet. Now find the number of acres: 87,120 square feet ÷ 43,560 square feet in an acre = 2 acres.

141. **a.** This is a multiplication problem. A quarter section contains 160 acres, so you must multiply: 160 × $1,850 = $296,000.

142. **c.** This is a subtraction problem. First, simplify the problem by dropping the word *million*. The problem then becomes P = 3,666 and M = 36. So P − M = 3,666 − 36 = 3,630. Now add the word *million* back, and the answer becomes 3,630 million.

143. **b.** 860 feet × 560 feet ÷ 43,560 square feet per acre = 11.06 acres

144. **b.** 30 men × 42 square feet = 1,260 square feet of space; 1,260 square feet ÷ 35 men = 36 square feet; 42 − 36 = 6, so each man will have 6 fewer square feet of space.

Set 10 (Page 23)

145. **d.** First, find the number of degrees that the temperature will increase; 57° − 53° = 4°. Since the temperature increases 1° every two hours, 4° × 2 hours is 8 hours. Add eight hours to 5 A.M. It will be one o'clock in the afternoon.

146. **b.** In a fractional exponent, the numerator (number on top) is the power, and the denominator (number on the bottom) is the root. Since 2 is the denominator, take the square root of 16. Since 4 × 4 = 16, 4 is the square root of 16. Then the numerator (power) is 1, so $4^1 = 4$.

147. **c.** The median of a group of numbers is found by arranging the numbers in ascending or descending order, and then finding the number in the middle of the set. First, arrange the numbers in order: 12, 14, 14, 16, 20, 22. Since there is an even number of numbers in the list, find the average of the two numbers that share the middle. In this case, the numbers in the middle are 14 and 16, and the average between them is 15.

148. **b.** When you are performing addition or subtraction with roots, the terms can be combined only if the radicands (numbers under the square root symbols) are the same. In this case, the radicands are both 2. Subtract the whole numbers in front of the square roots and keep the $\sqrt{2}$. Then, 16 − 4 = 12, so the final answer is $12\sqrt{2}$.

149. **a.** To reduce a square root, find any factors that are also perfect squares. In this case, 4 × 5 = 20, and 4 is a perfect square. Since $\sqrt{4} = 2$, then $\sqrt{20} = \sqrt{4} \times \sqrt{5} = 2\sqrt{5}$.

150. **c.** In order for a number to be divisible by six, it must be able to be divided by six without a remainder. A shortcut to check divisibility by six is to see if the number is divisible by both two and three. Since the 546 is even (ends in 6), the number is divisible by two. Since the sum of the digits is 5 + 4 + 6 = 15 and 15 is divisible by three, 546 is also divisible by three. Since 546 is divisible by both two and three, it is also divisible by six.

151. **c.** Take the words in order and substitute the letters and numbers. Patricia has (P =), four times (4 ×) the number of marbles Sean has (S). The statement becomes P = 4 × S, which is equal to P = 4S.

152. **c.** The correct answer is 10,447. It helps, if you are in a place where you can do so, to read the answer aloud; that way, you will likely catch any mistake. When writing numbers with four or more digits, begin at the right and separate the digits into groups of threes with commas.

153. **c.** The correct answer is 10,043,703. The millions place is the third group of numbers from the right. If any group of digits *except the first* has fewer than three digits, you must add a zero at the beginning of that group.

154. **c.** This is a multistep problem. First, figure out how many *dozen* eggs John's chickens produce per day: 480 ÷ 12 = 40 dozen eggs per day. Now figure out how much money John makes on eggs per day: $2 × 40 = $80 per day. Finally, figure out how much money John makes per week: $80 × 7 = $560 per week. The most common mistake for this problem is to forget to do the last step. It is important to read each problem carefully, so you won't skip a step.

155. **d.** When evaluating a negative exponent, take the reciprocal of the base evaluated with a positive exponent. In other words, since $5^2 = 25$, then 5^{-2} equals the reciprocal of 25, or $\frac{1}{25}$.

156. **b.** The vertical bars on either side of the expression tell you to find the *absolute value* of 4 − 15. To complete the question, find 4 − 15 to get −11. Then find the absolute value (the distance the number is away from zero on a number line) of −11, which is 11.

157. **d.** The production for Lubbock is equal to the total minus the other productions: 1,780 − 450 − 425 − 345 = 560.

158. **b.** The number 54 divided by 6 is 9.

159. **a.** The mean is equal to the sum of values divided by the number of values. Therefore, 8 raptors per day × 5 days = 40 raptors. The sum of the other six days is 34 raptors; 40 raptors − 34 raptors = 6 raptors.

160. **c.** The speed of the train, obtained from the table, is 60 miles per hour. Therefore, the distance from Chicago would be equal to $60t$. However, as the train moves on, the distance decreases from Los Angeles, so there must be a function of $-60t$ in the equation. At time $t = 0$, the distance is 2,000 miles, so the function is $2{,}000 - 60t$.

▶ Section 2—Fractions

Set 11 (Page 26)

161. b. Two of the four sections are shaded, so $\frac{2}{4}$ of the figure is shaded. Reduced, the answer is $\frac{1}{2}$.

162. b. Since there are three sections shaded out of a total of five sections, the part shaded is $\frac{3}{5}$.

163. c. To reduce a fraction to lowest terms, find the greatest common factor of the numerator and denominator. Then, divide both by this number. The greatest common factor of 42 and 56 is 14. Therefore, the correct answer is $\frac{3}{4}$.

164. d. Although there are other factors that 54 and 108 share, dividing both by the greatest common factor, 54, will result in the lowest terms in one step. The correct answer is $\frac{1}{2}$.

165. c. To add or subtract fractions, it is necessary for the fractions to share a common denominator. These two fractions have a common denominator of 9. Therefore, simply subtract the numerators and keep the denominator. The result is $\frac{3}{9}$, but in the answer choices it is expressed in lowest terms, $\frac{1}{3}$.

166. a. First, find the least common denominator—that is, convert all three fractions to sixteenths, and then add: $\frac{4}{16} + \frac{3}{16} + \frac{14}{16} = \frac{21}{16}$. Now reduce: $1\frac{5}{16}$.

167. d. To add mixed numbers, make sure the fractions have a common denominator. Then, add the numerators and keep the denominator: $\frac{1}{3} + \frac{1}{3} = \frac{2}{3}$. Then add the whole numbers: $5 + 7 + 2 = 14$. Combine these results to reach a final solution of $14\frac{2}{3}$.

168. a. Because the fractions within the equation have a common denominator, subtract the numerators and keep the denominator to get $\frac{7}{13}$. Then subtract the whole numbers to get 3. Combine the results for a final answer of $3\frac{7}{13}$.

169. b. When multiplying fractions, multiply the numerators by each other: $1 \times 4 = 4$. Then, multiply the denominators by each other: $5 \times 7 = 35$. The correct answer is $\frac{4}{35}$.

170. a. Before subtracting, you must convert both fractions to thirty-sixths: $\frac{15}{36} - \frac{14}{36} = \frac{1}{36}$.

171. a. First, find a common denominator for the fractions (21) and subtract: $\frac{14}{21} - \frac{12}{21} = \frac{2}{21}$. Then subtract the whole numbers: $78 - 10 = 68$. Combine for an answer of $68\frac{2}{21}$.

172. c. When dividing fractions, multiply the first fraction by the reciprocal of the divisor: $-\frac{5}{12} \times \frac{6}{1} = \frac{30}{12}$. Then, convert to a mixed number: $2\frac{1}{2}$. The negatives cancel each other out, so the answer is positive.

173. b. First, change any whole numbers or mixed numbers to improper fractions; 40 becomes $\frac{40}{1}$ and $2\frac{1}{2}$ becomes $\frac{5}{2}$. To divide fractions, multiply by the reciprocal of the number being divided by. Therefore, the expression becomes $\frac{40}{1} \times \frac{2}{5}$. By cross canceling common factors, this becomes $\frac{8}{1} \times \frac{2}{1}$, which is equal to $\frac{16}{1}$ or 16.

174. b. The correct answer is $1\frac{1}{6}$.

175. a. Again, in order to subtract the fractions, you must first find the least common denominator, which in this case is 40. The equation is then $\frac{35}{40} - \frac{24}{40} = \frac{11}{40}$.

176. a. Because you are adding numbers that do not have the same sign, subtract the fraction with the smaller absolute value from the fraction with the larger absolute value to find the result. There is already a common denominator of 5, so first subtract the fractional parts: $\frac{3}{5} - \frac{2}{5} = \frac{1}{5}$. Next, subtract the whole numbers: $4 - 1 = 3$. Therefore, the solution is $3\frac{1}{5}$.

Set 12 (Page 28)

177. b. To answer the problem you must first convert $\frac{1}{2}$ to $\frac{3}{6}$, and then add. The correct answer is $88\frac{1}{3}$.

178. c. First, subtract the whole numbers: $35 - 20 = 15$. Because the fractions have a common denominator, simply subtract the numerators and keep the denominator: $\frac{7}{9} - \frac{2}{9} = \frac{5}{9}$. Combine the whole number and fraction for the answer of $15\frac{5}{9}$.

179. c. First, find the least common denominator of the fractions, which is 9, and then add the fractions: $\frac{6}{9} + \frac{3}{9} = \frac{9}{9}$ or 1. Now add the whole numbers: $43 + 36 = 79$. Now add the results of the two equations: $1 + 79 = 80$.

180. d. The lowest common denominator is 21: $\frac{12}{21} - \frac{7}{21} = \frac{5}{21}$.

181. d. The lowest common denominator is 24: $\frac{20}{24} + \frac{9}{24} = \frac{29}{24}$. Convert to a mixed number: $1\frac{5}{24}$.

182. a. The product when multiplying reciprocals is always 1.

183. b. Because dividing two negatives results in a positive, the only possible answers are choices **b** and **c**. Choice **c** is incorrect, because it is the result of multiplying the fractions. Choice b, 2, is correct because it is the result of multiplying by the reciprocal of the divisor.

184. d. Convert the mixed number to an improper fraction and multiply: $\frac{38}{5} \times \frac{4}{9} = \frac{152}{45}$. Convert back to a mixed number: $3\frac{17}{45}$.

185. d. First, convert 2 to a fraction. Then, invert and multiply: $\frac{2}{6} \times \frac{1}{2} = \frac{2}{12}$. Reduce to lowest terms: $\frac{1}{6}$.

186. b. First, change the mixed numbers to improper fractions. That is, for each fraction, multiply the whole number by the denominator of the fraction, then add the numerator:

$2 \times 4 + 1 = 9$, so the fraction becomes $\frac{9}{4}$; $2 \times 7 + 4 = 18$, so the fraction becomes $\frac{18}{7}$. The equation then becomes $\frac{9}{4} \div \frac{18}{7}$. Now invert the second fraction and multiply the numerators and the denominators: $\frac{9}{4} \times \frac{7}{18} = \frac{63}{72}$. Reduced, the answer becomes $\frac{7}{8}$. (If you performed a multiplication operation instead of a division one, you got wrong answer choice **d**.)

187. c. Multiply across: $\frac{10}{216}$. Then reduce to lowest terms to get the answer: $\frac{5}{108}$.

188. b. Convert the fractions to mixed numbers: $1\frac{1}{2} = \frac{3}{2}$, and $1\frac{5}{13} = \frac{18}{13}$. Now invert the second fraction and multiply: $\frac{3}{2} \times \frac{13}{18} = \frac{39}{36}$ or $1\frac{3}{36}$. Now reduce: $1\frac{1}{12}$. (A common error, multiplying fractions instead of dividing, is shown in choice **c**.)

189. d. Properly converting the mixed numbers into improper fractions is the first step in finding the answer. Thus, $\frac{7}{3} \times \frac{15}{14} \times \frac{9}{5} = \frac{945}{210} = 4\frac{1}{2}$.

190. a. Multiply across to get the answer: $\frac{28}{45}$.

191. c. First, change $2\frac{1}{4}$ to an improper fraction: $2\frac{1}{4} = \frac{9}{4}$. Next, in order to divide by $\frac{2}{3}$, invert that fraction to $\frac{3}{2}$ and multiply: $\frac{9}{4} \times \frac{3}{2} = \frac{27}{8} = \frac{(24+3)}{8} = \frac{24}{8} + \frac{3}{8} = 3\frac{3}{8}$.

192. c. Invert the divisor and multiply: $\frac{4}{7} \times \frac{17}{8} = \frac{68}{56}$. Then, reduce and change to a mixed number: $1\frac{3}{14}$

Set 13 (Page 30)

193. b. To multiply a whole number by a fraction, change the whole number to a fraction by putting it over 1: $\frac{8}{1} \times \frac{1}{5} = \frac{8}{5}$. Change the improper fraction to a mixed number: $1\frac{3}{5}$.

194. c. The correct answer is $2\frac{4}{5}$.

195. d. Convert the mixed number to a fraction. Then invert it because it is the divisor. Multiply: $\frac{1}{6} \times \frac{8}{37} = \frac{8}{222}$. Reduce the fraction to lowest terms: $\frac{4}{111}$.

196. a. The correct answer is $28\frac{4}{7}$.

197. a. First, change each mixed number to an improper fraction: $1\frac{1}{2} = \frac{3}{2}$ and $2\frac{1}{4} = \frac{9}{4}$. The problem becomes $\frac{3}{2} \div \frac{9}{4}$. To divide fractions, multiply by the reciprocal of the number being divided by: $\frac{3}{2} \times \frac{4}{9} = \frac{12}{18} = \frac{2}{3}$.

198. a. First, change each of the mixed numbers to improper fractions. The expression becomes $\frac{7}{3} \times \frac{11}{2} \times \frac{3}{11}$. Cross cancel the factors of 3 and 11. Multiply across to get $\frac{7}{2}$, which is equal to $3\frac{1}{2}$.

199. b. The correct answer is 20.

200. a. The correct answer is $3\frac{11}{27}$.

201. c. Because there is a common denominator, the fractions can be added. The result is $-\frac{14}{7}$, which can also be represented as –2.

202. b. The correct answer in lowest terms is $\frac{4}{5}$.

203. a. The correct answer is 8. In fraction form, it is $\frac{40}{5}$.

204. a. The least common denominator for the two fractions is 18. Ignore the whole number 6 for a moment, and simply subtract the fractions: $\frac{4}{18} - \frac{3}{18} = \frac{1}{18}$. Now put back the 6 to find the correct answer: $6\frac{1}{18}$. (If you got answer choice **c**, you subtracted both the numerators and the denominators.)

205. c. The correct answer is $13\frac{29}{35}$.

206. a. The correct answer is $17\frac{8}{9}$.

207. d. Find the answer by changing the fractions to decimals: $\frac{1}{3} = 0.333$; $\frac{1}{4} = 0.25$; $\frac{2}{7} = 0.286$. The decimal 0.286, or $\frac{2}{7}$, is between the other two.

208. b. Divide the numerator by the denominator to find the whole number of the mixed number. The remainder, if any, becomes the numerator of the fraction: $55 \div 6 = 9$, remainder 1. The denominator stays the same. Therefore, the mixed number is $9\frac{1}{6}$.

Set 14 (Page 32)

209. d. Divide the numerator by the denominator to get the correct answer of 0.4.

210. a. To solve this problem, you must first convert all the fractions to the lowest common denominator, which is 24; $\frac{7}{8} = \frac{21}{24}$; $\frac{3}{4} = \frac{18}{24}$; $\frac{2}{3} = \frac{16}{24}$; $\frac{5}{6} = \frac{20}{24}$. The fraction with the largest numerator, $\frac{21}{24}$, has the greatest value.

211. b. Convert each fraction to the common denominator of 30: $\frac{3}{5} = \frac{18}{30}$; $\frac{8}{15} = \frac{16}{30}$; $\frac{17}{30}$ already has a denominator of 30; $\frac{2}{3} = \frac{20}{30}$. The fraction with the smallest numerator is $\frac{16}{30}$, which is equivalent to $\frac{8}{15}$.

212. d. To solve the problem, you must first find the common denominator, in this instance, 60. Then the fractions must be converted: $\frac{17}{20} = \frac{51}{60}$ (for choice **a**); $\frac{3}{4} = \frac{45}{60}$ (for choice **b**); $\frac{5}{6} = \frac{50}{60}$ (for choice **c**); and $\frac{7}{10} = \frac{42}{60}$ (for the correct answer, choice **d**).

213. b. Convert the mixed number $3\frac{7}{8}$ to the improper fraction $\frac{31}{8}$ and then invert to $\frac{8}{31}$.

214. c. Convert the mixed number to an improper fraction: $4\frac{3}{5} = \frac{23}{5}$. Then, invert the numerator and denominator: $\frac{5}{23}$.

215. d. Divide the top number by the bottom number: $160 \div 40 = 4$.

216. b. Divide the numerator by the denominator, and this number becomes the whole number of the mixed number. The remainder, if any, becomes the numerator of the fraction of the mixed number. So, $31 \div 3 = 10$ remainder 1. Therefore, the mixed number is $10\frac{1}{3}$.

217. a. Multiply the whole number by the fraction's denominator: $5 \times 2 = 10$. Add the fraction's numerator to the answer: $1 + 10 = 11$. Now place that answer over the fraction's denominator: $\frac{11}{2}$.

218. **d.** The first step is to convert each fraction to the least common denominator, which is 20. The problem becomes $\frac{12}{20} - \frac{5}{20}$. Subtract to get $\frac{7}{20}$.

219. **b.** In an improper fraction, the top number is greater than the bottom number, so $\frac{66}{22}$ is the correct answer.

220. **a.** Mario has finished $\frac{35}{45}$ of his test, which reduces to $\frac{7}{9}$, so he has $\frac{2}{9}$ of the test to go.

221. **a.** First, find the least common denominator of the two fractions, which is 6. Then add the fractions of the sandwich Joe got rid of: $\frac{3}{6}$ (which he gave to Ed) + $\frac{2}{6}$ (which he ate) = $\frac{5}{6}$. Now subtract the fraction from one whole sandwich ($1 = \frac{6}{6}$): $\frac{6}{6} - \frac{5}{6} = \frac{1}{6}$.

222. **a.** The total width of the three windows is 105 inches: $105 \times 2\frac{1}{2} = \frac{105}{1} \times \frac{5}{2} = \frac{525}{2} = 262\frac{1}{2}$.

223. **a.** Since Katie and her family ate $\frac{3}{4}$ of the pizza for dinner, $\frac{1}{4}$ of it is leftover. Katie ate $\frac{1}{2}$ of $\frac{1}{4}$ for lunch the next day: $\frac{1}{2} \times \frac{1}{4} = \frac{1}{8}$.

224. **c.** To subtract fractions, first convert to a common denominator, in this case, $\frac{25}{40} - \frac{24}{40} = \frac{1}{40}$.

Set 15 (Page 34)

225. **c.** To find out how many dozen cookies Hans can make, divide $5\frac{1}{2}$ by $\frac{2}{3}$. First, convert $5\frac{1}{2}$ to $\frac{11}{2}$, and then multiply by $\frac{3}{2}$, which is the same as dividing by $\frac{2}{3}$; $\frac{11}{2} \times \frac{3}{2} = \frac{33}{4}$, or $8\frac{1}{4}$ dozen.

226. **c.** Add the fractions and subtract the total from 2. The least common denominator is 24, so the fractions become $\frac{12}{24} + \frac{3}{24} + \frac{16}{24}$, which adds to $\frac{31}{24}$. Two pounds is equal to $\frac{48}{24}$, so $\frac{48}{24} - \frac{31}{24} = \frac{17}{24}$.

227. **a.** To multiply mixed numbers, convert to improper fractions, or $\frac{5}{4} \times \frac{11}{8} = \frac{55}{32}$, or $1\frac{23}{32}$ square inches.

228. **a.** Divide $6\frac{1}{2}$ pounds by eight people; $6\frac{1}{2} \div 8$ equals $\frac{13}{2} \div \frac{8}{1}$. To divide fractions, multiply by the reciprocal of the fraction being divided by: $\frac{13}{2} \times \frac{1}{8} = \frac{13}{16}$.

229. **c.** It is 36 linear feet around the perimeter of the room (9×4); $36 - 17\frac{3}{4} = \frac{73}{4}$ or $18\frac{1}{4}$.

230. **c.** Mixed numbers must be converted to fractions, and you must use the least common denominator of 8: $\frac{18}{8} + \frac{37}{8} + \frac{4}{8} = \frac{59}{8}$, which is $7\frac{3}{8}$ after it is reduced.

231. **d.** Since 28 of the 35 slices have been eaten, there are $35 - 28 = 7$ slices left. This means $\frac{7}{35}$, or $\frac{1}{5}$ of the loaf is left.

232. **a.** The common denominator is 24; $\frac{56}{24} - \frac{21}{24} = \frac{35}{24}$ or $1\frac{11}{24}$.

233. **c.** Convert both the cost and the length to fractions: $\frac{3}{4} \times \frac{22}{3} = \frac{66}{12}$ or $5\frac{1}{2}$, which is \$5.50.

234. **b.** The total number of pages assigned is 80; $\frac{1}{6} \times 80 = \frac{80}{6}$ or $13\frac{1}{3}$.

235. **d.** To subtract, convert to improper fractions, find a common denominator, and subtract the numerators: $\frac{11}{2} - \frac{10}{3} = \frac{33}{6} - \frac{20}{6} = \frac{13}{6}$ or $2\frac{1}{6}$.

236. **c.** To find the hourly wage, divide the total salary by the number of hours worked, or 331.01 divided by $39\frac{1}{2}$, or 39.5, which equals 8.38.

237. **d.** First, find how much of the cookie was eaten by adding the two fractions. After converting to the least common denominator, the amount eaten is $\frac{9}{21} + \frac{7}{21} = \frac{16}{21}$. This means $\frac{5}{21}$ of the cookie is left.

238. **b.** When subtracting mixed fractions, subtract the fractions first. Since 8 contains no fractions, convert to $7\frac{8}{8}$, and then subtract: in this case, $\frac{8}{8} - \frac{5}{8} = \frac{3}{8}$. Then subtract the whole numbers: in this case, $7 - 6 = 1$ (remember, 8 was converted to $7\frac{8}{8}$). Add the results: $1\frac{3}{8}$.

239. a. $\frac{3}{5}$ of 360 is figured as $\frac{3}{5} \times \frac{360}{1} = \frac{1,080}{5}$ or 216.

240. c. The least common denominator of the fractions is 20. When added together, the fraction part of each mixed number adds to $\frac{54}{20} = 2\frac{14}{20}$, which reduces to $2\frac{7}{10}$. The sum of the whole numbers is 36; $2\frac{7}{10} + 36 = 38\frac{7}{10}$.

Set 16 (Page 36)

241. c. The total number of items ordered is 36; the total received is 23. Therefore, Meryl has received 23 of 36 items or $\frac{23}{36}$; $\frac{23}{36}$ cannot be reduced.

242. b. To multiply fractions, convert to improper fractions: $7.75 = 7\frac{3}{4} = \frac{31}{4}$; $38\frac{1}{5} = \frac{191}{5}$. So, $\frac{31}{4} \times \frac{191}{5} = \frac{5,921}{20}$ or $296.05.

243. d. $\frac{1}{3}$ of $3\frac{1}{2}$ is expressed as $\frac{1}{3} \times 3\frac{1}{2}$, or $\frac{1}{3} \times \frac{7}{2} = \frac{7}{6}$ or $1\frac{1}{6}$.

244. a. To divide $20\frac{1}{3}$ by $1\frac{1}{2}$, first convert to improper fractions, or $\frac{61}{3}$ divided by $\frac{3}{2}$. To divide, invert the second fraction and multiply: $\frac{61}{3} \times \frac{2}{3} = \frac{122}{9} = 13.55$, or approximately $13\frac{1}{2}$ recipes.

245. c. There are two glasses out of eight left to drink, or $\frac{2}{8}$, which reduces to $\frac{1}{4}$.

246. a. There will be 100 questions on the test. Wendy has completed 25 of 100, or $\frac{1}{4}$ of the test.

247. c. To figure the necessary number of gallons, divide the number of miles by the miles per gallon: $\frac{117}{2}$ divided by $\frac{43}{3} = \frac{351}{86}$ or $4\frac{7}{86}$.

248. d. Felicia still needs 18 squares, or $\frac{18}{168}$, which can be reduced to $\frac{3}{28}$.

249. b. $14\frac{1}{2} \times 12\frac{1}{3}$, or $\frac{29}{2} \times \frac{37}{3} = \frac{1,073}{6}$ or 178.83 square feet. Two gallons of paint will cover 180 square feet.

250. b. The amount of meat loaf left is $\frac{2}{5} - (\frac{2}{5})(\frac{1}{4}) = \frac{2}{5} - \frac{2}{20}$. After you find the least common denominator, this becomes $\frac{8}{20} - \frac{2}{20} = \frac{6}{20}$, which reduces to $\frac{3}{10}$.

251. a. Millie has completed $\frac{16}{52}$, or $\frac{4}{13}$, of the total galleries.

252. c. First, convert $12.35 to a fraction ($12\frac{35}{100}$), and then convert it to an improper fraction ($\frac{1,235}{100}$) and reduce to $\frac{247}{20}$; $\frac{3}{4}$ of the workweek is 30 hours. To multiply a whole number by a fraction, convert the whole number to a fraction, $\frac{30}{1}$; $\frac{247}{20} \times \frac{30}{1} = \frac{7,410}{20}$ or $370\frac{1}{2}$ or $370.50.

253. b. $\frac{1}{2}$ of $\frac{1}{4}$ is expressed as $\frac{1}{2} \times \frac{1}{4}$ or $\frac{1}{8}$.

254. d. $\frac{1}{4} \times 2$ is expressed as $\frac{1}{4} \times \frac{2}{1} = \frac{2}{4}$, or $\frac{1}{2}$ teaspoon.

255. a. The total area of the lawn is 810 square yards (30×27). There is $\frac{1}{3}$ of the yard left to mow; $\frac{1}{3} \times \frac{810}{1} = \frac{810}{3}$ or 270.

256. d. The fraction of program time devoted to commercials is $\frac{6}{30}$, or $\frac{1}{5}$.

Set 17 (Page 38)

257. d. Let x equal the original weight. If $\frac{1}{8}$ of the original weight is equal to 15 pounds, then $\frac{1}{8} \times x = 15$. Multiply both sides by 8 to get $x = 120$ pounds.

258. d. To determine the number of bags each minute, divide the total bags by the total minutes, or $15\frac{1}{2}$ divided by 3, or $\frac{31}{2}$ divided by $\frac{3}{1}$, or $\frac{31}{2}$ times $\frac{1}{3}$, which is equal to $\frac{31}{6}$ or $5\frac{1}{6}$ bags per minute.

259. d. To determine $\frac{1}{4}$ of $5\frac{1}{3}$, multiply $\frac{1}{4}$ and $5\frac{1}{3}$. Change the mixed number to an improper fraction: $\frac{1}{4} \times \frac{16}{3} = \frac{16}{12} = \frac{4}{3} = 1\frac{1}{3}$.

260. **a.** This is a subtraction of mixed numbers problem. The common denominator is 8. Convert $\frac{1}{2}$ to $\frac{4}{8}$. Because $\frac{4}{8}$ is larger than $\frac{1}{8}$, you must borrow from the whole number 23. Then subtract: $22\frac{9}{8} - 6\frac{4}{8} = 16\frac{5}{8}$.

261. **a.** This is a subtraction problem. First, find the lowest common denominator, which is 12; $\frac{3}{4} = \frac{9}{12}$ and $\frac{2}{3} = \frac{8}{12}$. Then subtract: $\frac{9}{12} - \frac{8}{12} = \frac{1}{12}$.

262. **b.** To solve this problem, you must first convert yards to inches. There are 36 inches in one yard; $36 \times 3\frac{1}{3} = \frac{36}{1} \times \frac{10}{3} = \frac{360}{3} = 120$.

263. **b.** This is a subtraction of fractions problem. The common denominator is 8. Convert $22\frac{1}{4}$ to $22\frac{2}{8}$. Because $\frac{5}{8}$ is larger than $\frac{2}{8}$, you must borrow. Then, subtract: $21\frac{10}{8} - 17\frac{5}{8} = 4\frac{5}{8}$.

264. **c.** This is a simple subtraction problem with mixed numbers. First, find the lowest common denominator, which is 15; $\frac{4}{5} = \frac{12}{15}$ and $\frac{1}{3} = \frac{5}{15}$. Then subtract: $20\frac{12}{15} - 3\frac{5}{15} = 17\frac{7}{15}$.

265. **d.** First you must add the two fractions to determine what fraction of the total number of buses was in for maintenance and repair. The common denominator for $\frac{1}{6}$ and $\frac{1}{8}$ is 24, so $\frac{1}{6} + \frac{1}{8} = \frac{4}{24} + \frac{3}{24}$, or $\frac{7}{24}$. Next, divide 28 by $\frac{7}{24}$; $28 \div \frac{7}{24} = \frac{28}{1} \times \frac{24}{7} = \frac{672}{7} = 96$.

266. **d.** Mixed numbers must be converted to fractions, and you must use the least common denominator of 8: $\frac{18}{8} + \frac{37}{8} + \frac{4}{8} = \frac{59}{8}$, which is $7\frac{3}{8}$ after it is changed to a mixed number.

267. **b.** This is a problem involving addition of mixed numbers. First, find the common denominator, which is 6. Convert the fractions and add: $\frac{2}{6} + \frac{5}{6} + \frac{4}{6} = \frac{11}{6}$. Next, reduce the fraction: $\frac{11}{6} = 1\frac{5}{6}$. Finally, add the whole numbers and the mixed number: $2 + 1 + 2 + 1\frac{5}{6} = 6\frac{5}{6}$.

268. **b.** Solve this problem with the following equation: $37\frac{1}{2}$ hours $- 26\frac{1}{4}$ hours $= 11\frac{1}{4}$ hours.

269. **a.** Solve this problem with the following equations: $\frac{1}{2} + \frac{1}{2} + (5 \times \frac{1}{4}) = 1 + 1\frac{1}{4} = 2\frac{1}{4}$ miles.

270. **d.** This is an addition problem with fractions. Add the top numbers of the fractions: $\frac{1}{3} + \frac{1}{3} = \frac{2}{3}$. Then add the whole number: $1 + \frac{2}{3} = 1\frac{2}{3}$.

271. **b.** This is an addition problem with mixed numbers. First, add the fractions by finding the lowest common denominator; in this case, it is 40: $\frac{15}{40} + \frac{12}{40} + \frac{8}{40} = \frac{35}{40}$. Next, reduce the fraction: $\frac{35}{40} = \frac{7}{8}$. Then add the whole numbers: $7 + 6 + 5 = 18$. Finally, add the results: $18 + \frac{7}{8} = 18\frac{7}{8}$.

272. **d.** This is an addition problem. To add mixed numbers, first add the fractions: To do this, find the lowest common denominator; in this case, it is 20: $\frac{3}{4} = \frac{15}{20}$ and $\frac{3}{5} = \frac{12}{20}$. Now add: $\frac{15}{20} + \frac{12}{20} = \frac{27}{20}$. Next, reduce: $\frac{27}{20} = 1\frac{7}{20}$. Finally, add the whole numbers to get the result: $16 + 2 + 1\frac{7}{20} = 19\frac{7}{20}$.

Set 18 (Page 40)

273. **a.** In this problem, you must find the fraction. Paula has completed 15 of the 25 minutes, or $\frac{15}{25}$. Reduce the fraction: $\frac{15}{25} = \frac{3}{5}$.

274. **c.** This is a multiplication problem. To multiply a whole number by a mixed number, first convert the mixed number to a fraction: $1\frac{2}{5} = \frac{7}{5}$. Then, multiply: $\frac{7}{5} \times \frac{3}{1} = \frac{21}{5}$. Now reduce: $\frac{21}{5} = 4\frac{1}{5}$.

275. **c.** This is a division problem with mixed numbers. First, convert the mixed numbers to fractions: $33\frac{1}{2} = \frac{67}{2}$ and $5\frac{1}{4} = \frac{21}{4}$. Next, invert the second fraction and multiply: $\frac{67}{2} \times \frac{4}{21} = \frac{134}{21}$. Reduce to a mixed number: $\frac{134}{21} = 6\frac{8}{21}$. With this result, you know that only six glasses can be completely filled.

276. **d.** This is a division problem. First, change the mixed number to a fraction: $3\frac{1}{2} = \frac{7}{2}$. Next, invert $\frac{1}{8}$ and multiply: $\frac{7}{2} \times \frac{8}{1} = 28$.

277. **b.** This is a multiplication problem with fractions. Six minutes is $\frac{6}{60}$ of an hour, which is reduced to $\frac{1}{10}$; $2\frac{1}{2} = \frac{5}{2}$. Next, multiply: $\frac{1}{10} \times \frac{5}{2} = \frac{1}{4}$.

278. **d.** First, convert the $2\frac{1}{2}$ hours to minutes by multiplying $2\frac{1}{2} \times 60$ to get 150 minutes. Then divide the answer by 50, the number of questions: $150 \div 50 = 3$.

279. **b.** Use 35 for C: $F = (\frac{9}{5} \times 35) + 32$. Therefore, $F = 63 + 32$, or 95.

280. **b.** Because the answer is a fraction, the best way to solve the problem is to convert the known to a fraction: $\frac{10}{45}$ of the cylinder is full. By dividing both the numerator and the denominator by 5, you can reduce the fraction to $\frac{2}{9}$.

281. **c.** Solving this problem requires determining the circumference of the spool by multiplying $\frac{22}{7}$ by $3\frac{1}{2}$ ($\frac{7}{2}$). Divide the total (11) into 53. The answer is 4.8, so the hose will completely wrap only four times.

282. **a.** Convert Fahrenheit temperature to Celsius temperature using the given formula. Use F = 113; $C = \frac{5}{9}(113 - 32) = \frac{5}{9}(81) = 45$.

283. **b.** To solve this problem, find the number of gallons of water missing from each tank ($\frac{1}{3}$ from Tank A, $\frac{3}{5}$ from Tank B), and then multiply by the number of gallons each tank holds when full (555 for Tank A; 680 for Tank B): $\frac{1}{3} \times 555 = 185$ for Tank A; $\frac{3}{5} \times 680 = 408$ for Tank B. Now add the number of gallons missing from both tanks to get the number of gallons needed to fill them: $185 + 408 = 593$.

284. **a.** This is a multistep problem. First, determine how many pieces of pizza have been eaten. Eight pieces ($\frac{8}{9}$) of the first pizza have been eaten; 6 pieces ($\frac{6}{9}$) of the second pizza have been eaten; 7 pieces ($\frac{7}{9}$) of the third pizza have been eaten. Next, add the eaten pieces; $8 + 6 + 7 = 21$. Since there are 27 pieces of pizza in all, 6 pieces are left, or $\frac{6}{27}$. Reduce the fraction: $\frac{6}{27} = \frac{2}{9}$.

285. **c.** To solve the problem, you have to first convert the total time to minutes (for the correct answer choice **c**, this is 105 minutes), then multiply by 4 (420 minutes), and then convert the answer back to hours by dividing by 60 minutes to arrive at the final answer (7 hours). Or you can multiply $1\frac{3}{4}$ hours by 4 to arrive at the same answer.

286. **a.** In this problem, you must multiply a fraction by a whole number: $\frac{5}{6} \times \frac{5}{1} = \frac{25}{6}$. Convert to a mixed number: $\frac{25}{6} = 4\frac{1}{6}$.

287. **b.** This is a two-step problem involving both addition and division. First, add the two mixed numbers to find out how many ounces of jelly beans there are in all: $10\frac{1}{4} + 9\frac{1}{8} = 19\frac{3}{8}$. Convert the result to a fraction: $19\frac{3}{8} = \frac{155}{8}$. Next, to divide by 5, invert the whole number and multiply: $\frac{155}{8} \times \frac{1}{5} = \frac{31}{8}$. Convert to a mixed number: $\frac{31}{8} = 3\frac{7}{8}$.

288. **b.** If half the students are female, then you would expect half of the out-of-state students to be female. One half of $\frac{1}{12}$ equals $\frac{1}{2} \times \frac{1}{12}$, or $\frac{1}{24}$.

Set 19 (Page 42)

289. **b.** There are 60 minutes in one hour. Multiply $60 \times 7\frac{1}{6}$ by multiplying $60 \times 7 = 420$ and $60 \times \frac{1}{6} = 10$. Then add $420 + 10$ to get 430 minutes.

290. **a.** To solve this problem, you must convert $3\frac{1}{2}$ to $\frac{7}{2}$ and then divide $\frac{7}{2}$ by $\frac{1}{4}$. The answer, $\frac{28}{2}$, is then reduced to 14.

291. **c.** The simplest way to add these three numbers is first to add the whole numbers, and then add the fractions: $3 + 2 + 4 = 9$. Then, $\frac{1}{2} + \frac{3}{4} = \frac{2}{4} + \frac{3}{4} = \frac{5}{4}$, or $1\frac{1}{4}$. Then, $9 + 1\frac{1}{4} = 10\frac{1}{4}$.

292. **c.** This is a multiplication problem involving a whole number and a mixed number. There are 16 ounces in one pound, so you must multiply $16 \times 9\frac{1}{2}$. First, change the whole number and the mixed number to fractions: $\frac{16}{1} \times \frac{19}{2} = \frac{304}{2}$. Then convert: $\frac{304}{2} = 152$ ounces.

293. **b.** In this problem you must find the fraction: $\frac{7,500}{25,000}$. Next, reduce the fraction. The easiest way to reduce is to first eliminate zeros in the numerator and the denominator: $\frac{7,500}{25,000} = \frac{75}{250}$. Then, further reduce the fraction: $\frac{75}{250} = \frac{3}{10}$.

294. **d.** If two of five cars are foreign, three of five are domestic: $\frac{3}{5} \times 60$ cars $= 36$ cars.

295. **b.** $2\frac{1}{2}$ is equal to 2.5; $1\frac{1}{4}$ is equal to 1.25; 2.5×1.25 is equal to 3.125 or $3\frac{1}{8}$.

296. **c.** If 150 of the 350 seats are filled, then $\frac{150}{350}$ represents the part of the auditorium that is full. Divide each by the greatest common factor of 50 to reduce to $\frac{3}{7}$.

297. **b.** Use the formula beginning with the operation in parentheses: $98 - 32 = 66$. Then multiply 66 by $\frac{5}{9}$, first multiplying 66 by 5 to get 330; $330 \div 9 = 36.67$, which is rounded up to 36.7.

298. **c.** Substitute $C = 30$ into the formula provided; $F = \frac{9}{5}(30) + 32$; $F = 54 + 32$; $F = 86$.

299. **d.** This is a multiplication of fractions problem: $\frac{1}{3} \times \frac{210}{1} = 70$.

300. **d.** This is a two-step problem involving multiplication and simple subtraction. First, determine the amount of sand contained in the four trucks: $\frac{3}{4} \times \frac{4}{1} = \frac{12}{4}$. Reduce: $\frac{12}{4} = 3$. Then subtract: $3 - 2\frac{5}{6} = \frac{1}{6}$. There is $\frac{1}{6}$ ton more than is needed.

301. **b.** To solve the problem, you must first find the common denominator, in this instance, 24. Then the fractions must be converted: $\frac{1}{8} = \frac{3}{24}$; $\frac{1}{6} = \frac{4}{24}$; $\frac{3}{4} = \frac{18}{24}$. Add the values for first and second layers together: $\frac{3}{24} + \frac{4}{24} = \frac{7}{24}$, and then subtract the sum from the total thickness $(\frac{18}{24})$: $\frac{18}{24} - \frac{7}{24} = \frac{11}{24}$.

302. **c.** The recipe for 16 brownies calls for $\frac{2}{3}$ cup butter. An additional $\frac{1}{3}$ cup would make 8 more brownies, for a total of 24 brownies.

303. **a.** The recipe is for 16 brownies. Half of that, 8, would reduce the ingredients by half. Half of $1\frac{1}{2}$ cups of sugar is $\frac{3}{4}$ cup.

304. **c.** In this problem you must find the fraction: $\frac{6}{10}$ of the cake has been eaten, so $\frac{4}{10}$ of the cake is left. Now reduce the fraction: $\frac{4}{10} = \frac{2}{5}$.

Set 20 (Page 45)

305. **a.** Rounding to close numbers helps. This is approximately $100,000 \div 500,000 = 0.20$ or $\frac{1}{5}$.

306. **b.** Yellow beans + orange beans = 12. There are 30 total beans; $\frac{12}{30}$ is reduced to $\frac{2}{5}$.

307. **c.** In this problem, you must multiply a mixed number by a whole number. First, rewrite the whole number as a fraction: $5 = \frac{5}{1}$. Then rewrite the mixed number as a fraction: $6\frac{1}{2} = \frac{13}{2}$. Then multiply: $\frac{5}{1} \times \frac{13}{2} = \frac{65}{2}$. Finally, convert to a mixed number: $\frac{65}{2} = 32\frac{1}{2}$.

308. **c.** This is a division of fractions problem. First, change the whole number to a fraction: $6 = \frac{6}{1}$. Then invert the second fraction and multiply: $\frac{6}{1} \times \frac{4}{1} = 24$.

309. **d.** In this problem, you must multiply a fraction by a whole number. First, rewrite the whole number as a fraction: $8 = \frac{8}{1}$. Next, multiply: $\frac{8}{1} \times \frac{4}{5} = \frac{32}{5}$. Finally, convert to a mixed number: $\frac{32}{5} = 6\frac{2}{5}$.

310. **c.** This is a multiplication with mixed numbers problem. First, change both mixed numbers to fractions: $3\frac{1}{4} = \frac{13}{4}$; $1\frac{2}{3} = \frac{5}{3}$. Next, multiply the fractions: $\frac{13}{4} \times \frac{5}{3} = \frac{65}{12}$. Finally, change the result to a mixed number: $\frac{65}{12} = 5\frac{5}{12}$.

311. **b.** This is a division problem. First, change the mixed number to a fraction: $1\frac{1}{8} = \frac{9}{8}$. Invert the whole number 3 and multiply: $\frac{9}{8} \times \frac{1}{3} = \frac{3}{8}$.

312. **b.** This is a basic addition problem. First, change the fractions to decimals. Then, you might think of the problem this way: For Iris to get to Raoul's house, she must go 2.5 miles west to the Sunnydale Mall, then on west 1.5 miles to the QuikMart, and then on 4.5 miles to Raoul's house. So, 2.5 + 1.5 + 4.5 = 8.5 miles.

313. **a.** To find the area of the floor in square feet, multiply the length by the width: $9\frac{3}{4} \times 8\frac{1}{3}$. To multiply mixed numbers, first convert to improper fractions, or $\frac{39}{4} \times \frac{25}{3} = \frac{975}{12}$ or $81\frac{1}{4}$.

314. **b.** There are 12 inches in one foot. Change the mixed number to a decimal: 4.5. Now multiply: $4.5 \times 12 = 54$.

315. **a.** First, change the mixed numbers to decimals: $7\frac{1}{2} = 7.5$ and $4\frac{1}{4} = 4.25$. Now subtract: $7.5 - 4.25 = 3.25$. Now change the decimal back to a fraction: $3.25 = 3\frac{1}{4}$.

316. **c.** First, find the least common denominator of the fractions, which is 6. Then add: $9\frac{2}{6} + 8\frac{5}{6} = 17\frac{7}{6}$, or $18\frac{1}{6}$.

317. **d.** First, add the weight of Ralph's triplets: $4\frac{3}{8} + 3\frac{5}{6} + 4\frac{7}{8}$, or (after finding the least common denominator) $4\frac{9}{24} + 3\frac{20}{24} + 4\frac{21}{24} = 11\frac{50}{24}$, or

$13\frac{2}{24}$, or $13\frac{1}{12}$. Now find the weight of Harvey's twins: $7\frac{2}{6} + 9\frac{3}{10}$, or (after finding the least common denominator) $7\frac{10}{30} + 9\frac{9}{30} = 16\frac{19}{30}$. Now subtract: $16\frac{19}{30} - 13\frac{1}{12} = 16\frac{38}{60} - 13\frac{5}{60} = 3\frac{33}{60} = 3\frac{11}{20}$. So Harvey's twins outweigh Ralph's triplets by $3\frac{11}{20}$ pounds. (No further reduction of the fraction is possible.)

318. **d.** If she uses $\frac{3}{8}$ of a pound, then $\frac{5}{8}$ of a pound is left. The question asks for the number of ounces left, so convert one pound to 16 ounces. Then find $\frac{5}{8}$ of 16 ounces by multiplying: $\frac{5}{8} \times 16 = 10$ ounces.

319. **a.** To find the fraction, compare the down payment with the total cost of the car; $\frac{630}{6,300}$ reduces to $\frac{1}{10}$.

320. **c.** Change the information into a fraction: $\frac{2}{8}$. Now, reduce the fraction: $\frac{1}{4}$.

► Section 3—Decimals

Set 21 (Page 48)

321. **d.** The hundredth is the second digit to the right of the decimal point. Because the third decimal is 6, the second is rounded up to 4.

322. **d.** The farther to the right the nonzero digits are, the smaller the number. Forty three thousandths is greater than 43 ten-thousandths.

323. **c.** Common errors include choice **a**, subtracting instead of adding, or choice **d**, not lining up the decimal points correctly.

324. **c.** If you line up the decimal points properly, you don't even have to do the subtraction to see that none of the other answers is even close to the correct value.

325. **a.** To evaluate this type of division problem, you need to move each decimal point four spaces to the right, and then bring that point straight up into the dividend, or answer. If you got choice **c**, you divided 0.8758 by 2.9.

326. **b.** 195.6 ÷ 7.2 yields a repeating decimal, 27.1666666 . . ., which, rounded up to the nearest hundredth, is 27.17.

327. **c.** The other answers were subtracted without aligning the decimal point correctly. The correct placement of the decimal point is 510.8.

328. **a.** The correct answer is 49.98. Not lining up the decimal points correctly is the most common error in this type of equation.

329. **b.** The correct answer to this basic subtraction problem is 1.168.

330. **b.** After you multiply the digits, it is important to place the decimal point correctly in the answer: 148.32.

331. **d.** Because there are two decimal places in each of the numbers being multiplied, the product, 0.1904, will have four.

332. **d.** This is a mixed decimal, which included a whole number placed to the left of the decimal point. The zero is in the tenths place and the 4 is in the hundredths place: 5.04.

333. **b.** The correct answer is 1.

334. **a.** Because the digit in the hundredths place is less than 5, the tenths place would round down to 0.2.

335. **c.** The thousandths place is the third digit to the right of the decimal point, so 1.0086 is the correct answer.

336. **b.** The correct answer is 1.08.

Set 22 (Page 50)

337. **a.** It is important to align the decimal points, especially when adding vertically. The correct answer is 70.821.

338. **c.** There needs to be four digits to the right of the decimal point, so 0.0378 is the correct answer.

339. **a.** Divide as usual, using the long division algorithm. Then bring the decimal point up to get 3.965.

340. **b.** The correct answer is 261.667.

341. **a.** This is a simple addition problem to which 0.0793 is the correct answer.

342. **d.** A zero is assumed if there is no digit in the final place of a decimal number. After regrouping, the correct answer is 1.83.

343. **c.** Squaring 100 yields 10,000. Move the decimal point four places to the right in 4.32 to get the correct answer of 43,200.

344. **c.** The correct answer is 0.068.

345. **d.** There needs to be four digits to the right of the decimal point. Therefore, 122.2625 is the correct answer.

346. **c.** Ten times 10 times 10 is 1,000. One thousand times 7.25 is 7,250.

347. **d.** To multiply two numbers expressed in scientific notation, multiply the nonexponential terms (4.1 and 3.8) in the usual way. Then the exponential terms (10^{-2} and 10^4) are multiplied by adding their exponents. So $(4.1 \times 10^{-2})(3.8 \times 10^4) = (4.1 \times 3.8)(10^{-2} \times 10^4) = (15.58)(10^{-2+4}) = (15.58)(10^2) = 15.58 \times 10^2$. In order to express this result in scientific notation, you must move the decimal point one place to the left and add one to the exponent, resulting in 1.558×10^3.

348. **b.** To divide two numbers in scientific notation, you must divide the nonexponential terms (6.5 and 3.25) in the usual way, and then divide the exponential terms (10^{-6} and 10^{-3}) by subtracting the exponent of the bottom term from the exponent of the top term, so that you get $\frac{(6.5 \times 10^{-6})}{(3.25 \times 10^{-3})} = \frac{6.5}{3.25} \times \frac{10^{-6}}{10^{-3}} = 2 \times 10^{-6-(-3)} = 2 \times 10^{-6+3} = 2 \times 10^{-3}$.

349. **a.** The product is the answer when two numbers are multiplied. Therefore, 0.368 is the correct answer.

350. **d.** The correct answer to this simple addition problem is 1.268.

351. **c.** The decimal −0.16 is less than −0.06, the smallest number in the range.

Set 23 (Page 52)

352. **a.** The 8 is two places to the right of the decimal point, so the correct answer is eight hundredths.

353. **a.** The greatest value to the right of the decimal point can be determined by the tenths place. Choices **a**, **c**, and **d** all have a two in the tenths place. Choice **a** is correct because its values in the hundredths and thousandths places are greater than the other two possible answers.

354. **d.** Because there are zeros in both the tenths and hundredths place, 0.0097 is the lowest number of all the choices.

355. **b.** When you are rounding the hundredths place, it is necessary to look at the thousandths place. Since 8 is greater than 5, round up to 43.01.

356. **b.** First, add the three numbers to get a sum of 65.698. This number rounded to the nearest tenth is 65.7.

357. **a.** The sum of these three numbers is 42.6561.

358. **c.** Subtract the second number from the first. Then subtract the third number from that difference to get 2.87 as the correct answer.

359. **c.** The equation to be used is $(70T + 60T) = 325.75$, or $130T = 325.75$; $T = 2.5$.

360. **c.** Since $3.20 is the sale price, add $0.75 to $3.20 to find the original price. Line up the decimal points, and the result is $3.95.

361. **d.** To find the total, add the known amounts together: $1.42 + 1.89 = 3.31$ liters.

362. **a.** To subtract, write the first number as 17.80 and subtract 14.33. The answer is 3.47.

363. **b.** Add the numbers together to find the total of 34.47.

364. **d.** Add $15.75 to $67.98 and then subtract $27.58. The answer is $56.15.

365. **c.** First, add the cost of the wallpaper and carpet, which is $701.99. Subtract that from $768.56. The difference is $66.57.

366. **b.** One hundred cases is five times twenty cases, so the cost is 20 times $11.39, or $227.80.

367. **c.** To find the average, divide the total by the number of days: $87 ÷ 6 = $14.50 per day.

368. **b.** Eight times $23.65 is $189.20.

Set 24 (Page 54)

369. **c.** 5.96 divided by 4 equals 1.49.

370. **a.** There are twelve inches in one foot; 2.54 multiplied by 12 is 30.48.

371. **c.** In the first two hours, they are working together at the rate of one aisle per hour for a total of two aisles. In the next two hours, they work separately. Melinda works for two hours at 1.5 hours per aisle; 2 ÷ 1.5 = 1.33 aisles. Joaquin works for two hours at two hours per aisle. This is one aisle. The total is 4.33 aisles.

372. **a.** Calculate what Robin makes in an eight-hour day and then divide by 3, since Patrick makes $\frac{1}{3}$ of Robin's earnings; 21 × 8 = $168; $168 ÷ 3 = $56.

373. **b.** Forty-five minutes is equal to $\frac{3}{4}$ of an hour, so Reva will make only $\frac{3}{4}$ of her usual fee. Change $\frac{3}{4}$ to a decimal: 0.75. Now multiply: 10 × 0.75 = 7.5. Reva will make $7.50 today.

374. **c.** Change the fraction to a decimal: 0.25 (which is $\frac{1}{4}$ as fast, or 25% of Zelda's time). Now multiply: 35.25 × 0.25 = 8.8125, which can be rounded to 8.81.

375. **c.** This problem is done by dividing: 1.5 ÷ 600 = 0.0025 inch.

376. **b.** Since the time given is in minutes, convert to hours. There are 60 minutes in one hour, so $\frac{15}{60}$ = 0.25 hours. Use the formula *Distance = rate × time*, whereby the rate is 4.5 miles per hour and the time is 0.25 hours. Distance = (4.5)(0.25) = 1.125 miles. The fact that

Michael leaves at 7:32 A.M. is irrelevant to this question.

377. **b.** Kyra saves $60 + $130 + $70, which equals $260. In January, her employer contributes $60 × 0.1 = $6, and in April, her employer contributes $70 × 0.1 = $7. In March, her employer contributes only $10 (**not** $13), because $10 is the maximum employer contribution. The total in savings then is $260 + $6 + $7 + $10 = $283. (If you chose choice **c**, you forgot that the employer's contribution was a *maximum* of $10.)

378. **a.** This is a subtraction problem. Align the decimals and subtract: 428.99 − 399.99 = 28.99.

379. **c.** This is a multiplication problem with decimals: 2.5 × 2,000 = 5,000.

380. **c.** This is a division problem: 304.15 ÷ 38.5. Because there is one decimal point in 38.5, move the decimal point one place in both numbers: $\frac{3,041.5}{385}$ = 7.9.

381. **c.** This is a multiplication problem. To multiply a number by 1,000 quickly, move the decimal point three digits to the right—one digit for each zero. In this situation, because there are only two decimal places, add a zero.

382. **d.** This is a division problem. Divide 12.9 by 2 to get 6.45, and then add both numbers: 12.90 + 6.45 = 19.35.

383. **a.** This is a division problem. Because there are two decimal points in 1.25, move the decimal point two places in both numbers: $\frac{2,240}{125}$ = 17.92.

384. **b.** This is a simple subtraction problem. Be sure to align the decimal points: 99.0 − 97.2 = 1.8.

Set 25 (Page 56)

385. **b.** Both addition and subtraction are required to solve this problem. First, add the amounts of the three purchases together: $12.90 + 0.45 + 0.88 = 14.23$. Next, subtract this amount from 40: $40.00 - 14.23 = 25.77$.

386. **d.** This problem requires both multiplication and addition. First, multiply 2.12 by 1.5 to find the price of the cheddar cheese: $2.12 \times 1.5 = 3.18$. Then add: $3.18 + 2.34 = 5.52$.

387. **b.** This problem requires both addition and subtraction. First, add the three lengths of string: $5.8 + 3.2 + 4.4 = 13.4$. Then subtract the answer from 100, making sure to align the decimal points: $100.0 - 13.4 = 86.6$.

388. **c.** This is a simple division problem; $\frac{2.00}{3} = 0.666$. Because 6 is higher than 5, round up to 7.

389. **a.** This is a simple addition problem. Line up the decimals in a column so that the decimal points are aligned: $9.4 + 18.9 + 22.7 = 51$.

390. **b.** This is a multiplication problem with decimals. Manny spends $1.10 each way and makes 10 trips each week: $1.10 \times 10 = 11.00$.

391. **b.** Divide 0.75 by 0.39 to get approximately 1.923 centimeters.

392. **c.** $5.133 \times 10^{-6} = 5.133 \times 0.000001 = 0.000005133$. This is the same as simply moving the decimal point to the left six places.

393. **a.** Add the four numbers. The answer is $23.96.

394. **b.** Add all of the money gifts as well as what Hal earned. The total is $4,772.56. Subtract this number from the cost of the car. The remainder is $227.39.

395. **c.** Subtract 2.75 from 10 by adding a decimal and zeros to the 10: $10.00 - 2.75 = 7.25$. Don't forget to line up the decimals.

396. **d.** Remember to ignore the decimals when multiplying $8,245 \times 92 = 758,540$. Then, total the number of decimal points in the two numbers you multiplied (four) and place the decimal point four places from the right in the answer: 75.8540.

397. **b.** Divide 517.6 by 23.92. Don't forget to move both decimals two spaces to the right (add a zero to 517.6); move the decimal up directly and divide. The answer is 21.6387. Round up to 21.64.

398. **d.** To find the difference, subtract 22.59 from 23.7. To keep the decimal placement clear, remember to add a zero to 23.7.

399. **a.** Add the three amounts, adding zeros where necessary. The total is 11.546, rounded down to 11.5 pounds.

400. **d.** This is a multiplication problem. Multiply the approximate words per minute times the number of minutes: $41.46 \times 8 = 331.68$.

Set 26 (Page 58)

401. **a.** Add all the candy Ingrid distributed or consumed and subtract that number from 7.5. The result is 5.44 pounds.

402. **c.** Add all the distances together. The sum is 16.32.

403. **b.** The decimal 20.34 divided by 4.75 equals 4.282. Ken can cover four chairs.

404. **d.** The total balance before expenditures is $2,045.38. The total expenditure is $893.14. Subtracted, the total is $1,152.24.

405. **d.** Multiply the length by the width to get 98.4375. Since the hundredths place is two places to the right of the decimal, the rounded answer is 98.44 square inches.

406. **b.** Five pounds of flour divided by 0.75 equals 6.6666. . . . Michael can make six cakes.

407. **d.** Add all the distances and divide by the number of days she walked; 4.38 divided by 4 = 1.095.

408. **c.** Add the eaten amounts and subtract from the total: 12.600 − 6.655 = 5.945 pounds of clams are left. (If you chose choice **d**, you forgot the last step of the problem.)

409. **a.** Solve 2.4 − 1.025, being careful to line up the decimal points. The answer is 1.375 million dollars.

410. **b.** Divide 76 by 2.89. This equals about 26.29 square feet, which rounds to 26 square feet.

411. **b.** Subtract the hours actually worked from the hours usually worked; the number of hours is 4.75.

412. **a.** Add the three amounts together. The total is 1.63 acres.

413. **d.** To find the area, multiply the length by the width. The area is 432.9 square feet.

414. **a.** The total of ads and previews is 11.3 minutes. Two hours is 120 minutes; 120 − 11.3 = 108.7.

415. **c.** Ten yards divided by 0.65 equals $15.\overline{384615}$ (repeating). Tommy can make 15 hats.

416. **b.** Add the four amounts. The total is 28.13 hours.

Set 27 (Page 61)

417. **c.** The distance around this section is the perimeter of the rectangle. This is found by adding the two known dimensions and multiplying by 2, since there are two pairs of sides that are the same: 2(16.25 + 20.25) = 2(36.5) = 73 feet.

418. **c.** 8.6 million minus 7.9 million equals 0.7 million.

419. **b.** $1.13 multiplied by 100 equals $113.00. Remember, a shortcut for multiplying fractions by 10, 100, 1,000, etc. is simply to move the decimal to the right one space for each zero.

420. **d.** The four distances added together equal 5.39 miles.

421. **b.** Fifteen minutes is $\frac{1}{4}$, or 0.25, of an hour; 0.25 of 46.75 is 0.25 × 46.75 = 11.6875.

422. **a.** Multiply 46.75 by 3.80, which equals 177.65.

423. **c.** The ratio of boys to total students is 3:4, or $\frac{3}{4}$, which is equal to 0.75; 0.75 × 28 = 21.

424. **a.** This is a division problem with decimals; 4,446 ÷ 1.17 = 3,800.

425. **b.** You must divide two decimals: 20.32 ÷ 2.54. First, move each number over two decimal places: 2,032 ÷ 254 = 8.

426. **c.** This is a four-step problem. First, determine how much she earns in one eight-hour day: 8 × $12.50 = $100. Next, subtract $100 from $137.50 to find how much overtime she earned: $137.50 − $100 = $37.50. Next, to find out how much her hourly overtime pay is, multiply 1.5 × 12.50, which is 18.75. To find out how many overtime hours she worked, divide: 37.50 ÷ 18.75 = 2. Add these two hours to her regular eight hours for a total of 10 hours.

427. **b.** Solving this problem requires converting 15 minutes to 0.25 hour, which is the time, and then using the formula $d = rt$: 62 miles per hour × 0.25 hour = 15.5 miles.

ANSWERS

428. **d.** Distance traveled is equal to velocity (or speed) multiplied by time. Therefore, $3.00 \times (10^8) \frac{meters}{second} \times 2{,}000$ seconds $= 6.00 \times 10^{11}$ meters.

429. **a.** First it is necessary to convert centimeters to inches. To do this for choice **a**, multiply 100 centimeters (1 meter) by 0.39 inches, yielding 39 inches. For choice **b**, 1 yard is 36 inches. For choice **d**, multiply 85 centimeters by 0.39 inches, yielding 33.15 inches. Choice **a**, 39 inches, is the longest.

430. **b.** To solve this problem, divide the number of pounds in the weight of the cat (8.5) by the number of pounds in a kilogram (2.2); 8.5 divided by 2.2 equals approximately 3.9 kilograms.

431. **b.** This is a multiplication problem: $1.25 times 40 is $50.

432. **b.** To find the total cost, add the four amounts together. Be sure to line up the decimal points. The total is $46.06.

Set 28 (Page 63)

433. **c.** The answer is arrived at by first dividing 175 by 45. Since the answer is 3.89, not a whole number, the firefighter needs four sections of hose. Three sections of hose would be too short.

434. **c.** First, convert 10 gallons into quarts. Since there are four quarts in one gallon, there are 40 quarts in 10 gallons. Now divide 40 quarts by 1.06, since one liter is equal to 1.06 quarts. Forty divided by 1.06 is approximately equal to 37.74, or about 38 liters.

435. **c.** To solve the problem, take the weight of one gallon of water (8.35) and multiply it by the

number of gallons (25): $8.35 \times 25 = 208.75$. Now round to the nearest unit, which is 209.

436. **a.** Because there are three at $0.99 and two at $3.49, the sum of the two numbers minus $3.49 will give the cost.

437. **d.** 2,200(0.07) equals $154; 1,400(0.04) equals $56; 3,100(0.07) equals $217; 900(0.03) equals $27. Therefore, $154 + $56 + $217 + $27 = $454. The other recycler offers only $440.

438. **a.** To solve the problem, multiply 3.5 pounds by 7, the number of days in one week.

439. **b.** The solution is simply the ratio of the rates of work, which is 15.25:12.5, or $\frac{15.25}{12.5}$ or 1.22. (To check your work multiply: 12.5 units $\times$ 1.22 hours = 15.25.)

440. **d.** $12.50 per hour $\times$ 8.5 hours per day $\times$ 5 days per week is $531.25.

441. **d.** Three inches every 2 hours = 1.5 inches per hour $\times$ 5 hours = 7.5 inches.

442. **d.** This can be most quickly and easily solved by estimating, that is, by rounding the numbers to the nearest ten cents: two hamburgers at $3 = $6; $6 + one cheeseburger at $3.40 = $9.40; two chicken sandwiches at $4 = $8. Then, $8 + one grilled cheese at $2 = $10. $9.40 + $10 = $19.40. Therefore, the nearest and most reasonable answer would be choice **d**, $19.10.

443. **c.** This is a simple multiplication problem that is solved by multiplying 35 times 8.2 for a total of 287.

444. **d.** This problem is solved by dividing 60 (the time) by 0.75 (the rate), which gives 80 words.

445. **c.** To find the answer, solve this equation: ($2.24 − $2.08) $\times$ 2 = $0.32.

446. **d.** First, find the total price of the pencils: 24 pencils $\times$ $0.05 = $1.20. Then find the total price of the paper: 3.5 reams $\times$ $7.50 per ream

= $26.25. Next, add the two totals together: $1.20 + 26.25 = $27.45.

447. c. This is a two-step multiplication problem. To find out how many heartbeats there would be in one hour, you must multiply 72 by 60 minutes, and then multiply this result, 4,320, by 6.5 hours.

448. c. Multiply $7.20 by 2.5 to get $18. Be sure to move the decimal to the left the correct number of spaces in your answer.

Set 29 (Page 66)

449. c. This is a simple addition problem. Add 1.6 and 1.5, keeping the decimal points aligned: 1.6 + 1.5 = 3.1.

450. b. This is a division problem: 25.8 ÷ 3 = 8.6. Move the decimal point straight up into the quotient.

451. c. You arrive at this answer by knowing that 254 is 100 times 2.54. To multiply by 100, move the decimal point two digits to the right.

452. a. This is an addition problem. Add the three numbers together, making sure the decimal points are aligned.

453. d. This is a multiplication problem: 35.2 × 71 = 2,499.2. There is only one decimal point, so you will count off only one place from the right.

454. b. This is a multiplication problem. First, multiply 279 by 89. Then, because there are four decimal places, count off four places from the right. Your answer should be 2.4831. Because the 3 in the thousandths place is less than 5, round to 2.48.

455. a. This is a basic subtraction problem. Line up the decimals and subtract: 2,354.82 − 867.59 = 1,487.23.

456. b. This is a division problem: 13.5 ÷ 4 = 3.375. Move the decimal straight up into the quotient.

457. b. This is a simple subtraction problem. Line up the decimals and subtract: 46.1 − 40.6 = 5.5.

458. d. This is an addition problem. To add these three decimals, line them up in a column so that their decimal points are aligned: 52.50 + 47.99 + 49.32 = 149.81. Move the decimal point directly down into the answer.

459. c. This is a basic addition problem. Be sure to align the decimal points before you add 68.8 + 0.6 = 69.4.

460. b. This is an addition problem. Arrange the numbers in a column so that the decimal points are aligned: 2.345 + 0.0005 = 2.3455.

461. c. This is a multiplication problem. Multiply 3.25 times 1.06. Be sure to count four decimal places from the right: 3.25 × 1.06 = 3.445.

462. c. This is an addition problem. Arrange the three numbers in a column so that the decimal points are aligned: 13.95 + 5.70 + 4.89 = 24.54.

463. c. This is an addition problem with decimals. Add the four numbers together to arrive at the answer, which is $1,772.10.

464. d. This is a two-step multiplication problem. First, find out how long it would take for both Bart and Sam to do the job: 0.67 × 5 = 3.35. Then, multiply your answer by 2 because it will take Bart twice as long to complete the job alone: 3.35 × 2 = 6.7

Set 30 (Page 68)

465. a. This is a simple subtraction problem. Align the decimal points and subtract: 91,222.30 − 84,493.26 = 6,729.04.

466. **a.** This is a division problem. Because there are two decimal digits, move the decimal point two places to the right in both numbers. This means you must tack a zero on to the end of 2,328. Then divide: $23,280 \div 375 = 62.08$.

467. **a.** This is a three-step problem that involves multiplication, addition, and subtraction. First, to determine the cost of the shrimp, multiply 3.16 by 4: $3.16 \times 4 = 12.64$. Then add the price of both the shrimp and the beef: $12.64 + 12.84 = 25.48$. Finally, subtract to find out how much money is left: $100.00 - 25.48 = 74.52$.

468. **c.** This is a two-step problem involving multiplication and division. First, determine the length of the pipes in inches by multiplying: $15.4 \times 3 = 46.2$. Next, divide to determine the length in feet: $46.2 \div 12 = 3.85$. Because there are no decimal points in 12, you can move the decimal point in 46.2 straight up into the quotient.

469. **c.** This is a multiplication problem. Be sure to count four decimal places from the right in your answer: $28.571 \times 12.1 = 345.7091$.

470. **b.** This is an addition problem. Arrange the three numbers in a column and be sure that the decimal points are aligned. Add: $0.923 + 0.029 + 0.1153 = 1.0673$.

471. **c.** This is a two-step problem involving both addition and division. First, arrange the three numbers in a column, keeping the decimal points aligned. Add: $113.9 + 106.7 + 122 = 342.6$. Next, divide your answer by 3: $342.6 \div 3 = 114.2$.

472. **b.** This is an addition problem. Be sure the decimal points are aligned before you add: $0.724 + 0.0076 = 0.7316$.

473. **d.** This problem involves two steps: addition and subtraction. Add to determine the amount of money Michael has: $20.00 + 5.00 + 1.29 = 26.29$. Then, subtract the amount of the ice cream: $26.29 - 4.89 = 21.40$.

474. **c.** This is a two-step problem. First, multiply to determine how many pounds of beef were contained in the eight packages: $0.75 \times 8 = 6$. Then add: $6 + 0.04 = 6.04$.

475. **d.** This is a two-step multiplication problem. First, multiply: $5 \times 2 = 10$, which is the number of trips Jacqueline drives to get to work and back. Then multiply 19.85 by 10 by simply moving the decimal one place to the right.

476. **a.** This is a simple subtraction problem: $42.09 - 6.25 = 35.84$.

477. **c.** This is a three-step problem. First, multiply to determine the amount Antoine earned for the first 40 hours he worked: $40 \times 8.3 = 332$. Next, multiply to determine his hourly wage for his overtime hours: $(1.5 \times 8.3)4 = 49.8$. Finally, add the two amounts: $332 + 49.8 = 381.8$.

478. **c.** This is a two-step problem involving both addition and subtraction. First add: $93.6 + 0.8 = 94.4$. Then subtract: $94.4 - 11.6 = 82.8$.

479. **a.** This is a two-step multiplication problem. First, multiply to find out how many weeks there are in six months: $6 \times 4.3 = 25.8$. Then, multiply to find out how much is saved: $\$40 \times 25.8 = \$1,032$.

480. **a.** This is a subtraction problem. Be sure to align the decimal points: $6.32 - 6.099 = 0.221$.

▶ Section 4—Percentages

Set 31 (Page 72)

481. **c.** To change a percent to a decimal, either divide the number by 100 and remove the percent sign, or move the decimal point two places to the left and remove the percent sign. Therefore, 0.06 is the correct answer.

482. **c.** The percent sign has been dropped, and the decimal point has moved two places to the left, so 0.09 is the correct answer.

483. **b.** The correct answer is 0.37.

484. **c.** The correct answer is 0.0304.

485. **c.** One hundred percent equals one whole. Therefore, 500% equals 5.0 wholes.

486. **b.** To convert a decimal to a percent, multiply the decimal by 100, or move the decimal point two places to the right. Therefore, 6.0% is the correct answer.

487. **c.** Convert the mixed number to a decimal to get 4.20%.

488. **d.** The decimal point has been moved two places to the left to get 0.2.

489. **c.** Convert the fraction to a decimal and keep the percent sign: $\frac{1}{5}\% = 0.20\%$.

490. **a.** Change the fraction to a decimal and move the decimal two places to the right. Then add the percent symbol to get 20%.

491. **a.** The decimal point has been moved two places to the right to get 66%.

492. **a.** $25\% = \frac{25}{100}$

493. **b.** First, change the percentage to a fraction by placing $\frac{625}{100}$. Then, change it to a mixed number and reduce to lowest terms: $6\frac{1}{4}$.

494. **c.** The fraction $\frac{32}{100}$ reduced to lowest terms is $\frac{8}{25}$.

495. **a.** Change the percent to a decimal to get 0.70. Then, multiply: $0.70 \times 600 = 420$.

496. **b.** Change the percent to a decimal to get 0.40. Then, multiply: $0.40 \times 240 = 96$.

Set 32 (Page 74)

497. **d.** The correct answer is 405.45.

498. **c.** Convert the percent to a decimal and multiply: $2.0 \times 40 = 80$.

499. **b.** The correct answer is 11.7.

500. **c.** The correct answer is 0.72.

501. **a.** The correct answer is 126.24.

502. **d.** A percentage is a portion of 100 where $x\% = \frac{x}{100}$. So the equation is $\frac{x}{100} = \frac{216}{12,000}$. Cross multiply: $12,000x = 216 \times 100$. Simplify: $x = \frac{21,600}{12,000}$. Thus, $x = 1.8\%$.

503. **d.** The correct answer is 175%.

504. **a.** To solve the problem, first change the percent to a decimal: $0.0425 \times 574 = 24.395$. Then, round to the nearest tenth: 24.4.

505. **b.** $62.5\% = \frac{62.5}{100}$. You should multiply both the numerator and denominator by 10 to move the decimal point, resulting in $\frac{625}{1,000}$, and then factor both the numerator and denominator to find out how far you can reduce the fraction: $\frac{625}{1,000} = \frac{(5)(5)(5)(5)}{(5)(5)(5)(8)}$. If you cancel the three 5s that are in both the numerator and denominator, you will get $\frac{5}{8}$.

506. **c.** First, divide the numerator by the denominator. Then, multiply by 100, or move the decimal point two places to the right and add the percent sign: $6 \div 80 = 0.075$; $0.075 \times 100 = 7.5\%$.

507. b. A quick way to figure a 20% tip is to calculate 10% of the total and then double that amount; 10% of $16.00 is $1.60; $1.60 multiplied by 2 is $3.20. Another method is to multiply 0.20 × $16 = $3.20 to get the solution.

508. c. A quick way to figure a 15% tip is to calculate 10% of the total, and then add half that amount to the 10%. In this case, 10% of $24 is $2.40, and half of $2.40 is $1.20; $2.40 + $1.20 = $3.60. Another method is to multiply 0.15 × $24 = $3.60 to get the solution.

509. d. The correct answer is 9.576.

510. b. The correct answer is 0.12.

511. c. The correct answer is 2.52.

512. a. To change a percent to a decimal, move the decimal point two places to the left and drop the percent sign: 0.6 and $\frac{6}{10}$.

Set 33 (Page 76)

513. d. Change the percent to a decimal and divide: $18 \div 0.25 = 72$.

514. b. The root *cent* means *100* (think of the word *century*), so the word *percent* literally means "per 100 parts." Thus 25% means 25 out of 100, which can also be expressed as a ratio: 25:100.

515. c. A ratio is a comparison of two numbers.

516. b. The correct answer is $\frac{34}{100}$.

517. d. The fraction bar represents the operation of division. To evaluate a fraction, always divide the top number (numerator) by the bottom number (denominator).

518. c. The equation to use is $\frac{x}{100} = \frac{12}{50}$. Cross multiply to get: $50x = (12)(100)$, or $x = \frac{1,200}{50}$. So $x = 24$, which means 12 is 24% of 50.

519. a. Change the percent to a decimal and divide: $21.12 \div 0.12 = 176$.

520. b. Divide the numerator by the denominator, and move the decimal point two places to the right. Then, add the percent sign: $5 \div 20 = 0.25$; 25%.

521. c. It is easy to mistake 0.8 for 8%, so you must remember: To get a percent, you must move the decimal point *two* spaces to the right. Since there is nothing on the right of 0.8, you must add a zero, then tack on the percent sign: 80%.

522. a. Remember that $\frac{1}{4} = 1 \div 4$ or 0.25; and 100% = 1. Therefore, $1 \times 0.25 = 0.25 = 25\%$.

523. b. There has been an increase in price of $3; $3 \div $50 = 0.06. This is an increase of 0.06, or 6%.

524. b. First, ask the question: "$42 is 12% of what number?" Change the percent to a decimal and divide $42 by 0.12. This is equal to $350, which represents the total weekly earnings. Now subtract the amount she spent on DVDs from $350: $350 − $42 = $308. She deposited $308 into her savings account.

525. b. First, change the percent to a decimal: $3\frac{1}{4}\% = 3.25\% = 0.0325$. Now multiply: $30,600 \times 0.0325 = 994.5$. Finally, add: $30,600 + $994.50 = $31,594.50 for Yetta's current salary.

526. d. The basic cable service fee of $15 is 75% of $20.

527. a. This is a multiplication problem involving a percent. Because 30% is equivalent to the decimal 0.3, simply multiply the whole number by the decimal: $0.3 \times 340 = 102$.

528. b. The problem asks what percent of 250 is 10. Since $x\% = \frac{x}{100}$, the equation is $\frac{x}{100} = \frac{10}{250}$. Cross multiply: $250x = (10)(100)$. Simplify: $x = \frac{1,000}{250}$ or $x = 4$. Thus 4% of the senior class received full scholarships.

Set 34 (Page 78)

529. **c.** This is a multiplication problem. Change 20% to a decimal and multiply: $13.85 \times 0.2 = 2.77$.

530. **c.** This percent problem involves finding the whole when the percent is given; 280,000 is 150% of last month's attendance. Convert 150% to a decimal; 150% = 1.5; 280,000 = 1.5 × LMA. Next, divide: $280,000 \div 1.5 = 186,666.6666\ldots$. Round up to the nearest whole number: 186,667.

531. **a.** This is a two-step problem. First, find the amount of profit. Convert the percent to a decimal and multiply. $70,000 \times 0.18 = 12,600$. Next, add the result to the original price: $70,000 + 12,600 = 82,600$.

532. **d.** To find the percent of decrease, first calculate the amount of the decrease: $1.00 - 0.95 = 0.05$. Set up the formula to solve for percent. Since $x\% = \frac{x}{100}$, the equation is $\frac{x}{100} = \frac{0.05}{1.00}$. Cross multiply: $(1.00)(x) = (0.05)(100)$. Simplify: $x = 5$. There is a 5% decrease.

533. **b.** To solve this problem, convert the percent to a decimal: $1\frac{1}{2}\% = 0.015$. Then multiply: $48.50 \times 0.015 = 0.7275$. Round up to 0.73.

534. **c.** Let x equal the percent that 24 is of 480, and set up the proportion $\frac{24}{480} = \frac{x}{100}$. Cross multiply to get $480x = 2,400$. Divide both sides of the equation by 480 to get $x = 5$. Therefore, 24 is 5% of 480.

535. **b.** Set up the problem this way: $\frac{x}{90} = \frac{10}{100}$. If you're getting good at percentages, you may just see the answer, but if you don't, cross multiply: $10 \times 90 = 100x$. Solve: $x = 9$.

536. **a.** The equation is $\frac{3}{16} = \frac{x}{100}$. Cross multiply: $3 \times 100 = 16x$. Solve for x by dividing by 16: $x = 18.75$.

537. **c.** To figure 19% of $423, use the equation $\frac{x}{423} = \frac{19}{100}$. Cross multiply: $100x = 423 \times 19$. Solving for x gives you 80.37. Subtract that answer from $423 to get $342.63.

538. **d.** First, find the percentage of men by subtracting 60% from 100%, which is 40%. Then, set up a proportion showing 40% out of 100% is equal to 250 men out of x total students; $\frac{250}{x} = \frac{40}{100}$. Cross multiply to get $40x = 25,000$. Divide both sides of the equation by 40 to get $x = 625$ total students.

539. **b.** To find 80% of $7.50, change 80% to decimal form and multiply: $0.80 \times 7.50 = \$6.00$.

540. **a.** Let x equal the percent of increase, and use the proportion $\frac{\text{difference}}{\text{original amount}} = \frac{x}{100}$. The difference is found by subtracting $8,990 - 7,865$, which is equal to 1,125. The proportion then becomes $\frac{1,125}{7,865} = \frac{x}{100}$. Cross multiply to get $7,865x = 112,500$. Divide each side of the equation by 7,865 to get $x \approx 14.30\%$, which rounds to 14%.

541. **d.** $10,000 \times 4.5 = 100x$; $x = 450$.

542. **b.** You can't just take 25% off the original price, because the 10% discount after three years of service is taken off the price that has already been reduced by 15%. Figure the problem in two steps: After the 15% discount the price is $71.83. Ninety percent of that—subtracting 10%—is $64.65.

543. **b.** This is a two-step problem. First, determine what percent of the trees are not oaks by subtracting: $100\% - 32\% = 68\%$. Change 68% to a decimal and multiply: $0.68 \times 400 = 272$.

544. **b.** In this problem, you must find the whole when a percent is given: 3,000 = 15% of Annual Budget. Change to a decimal: 3,000 = 0.15 × AB. Solve the equation: 3,000 = 0.15 × 20,000.

Set 35 (Page 80)

545. **d.** To find what percent one number is of another, first write out an equation. Since $x\% = \frac{x}{100}$ the equation is $\frac{x}{100} = \frac{420}{1,200}$. Cross multiply: $1,200x = (420)(100)$. Simplify: $x = \frac{42,000}{1,200}$. Thus $x = 35$, which means 35% of the videos are comedies.

546. **c.** To find the percent change, first determine the original weight: 65.1 − 5.1 = 60. Recalling that $x\% = \frac{x}{100}$, the equation becomes $\frac{x}{100} = \frac{5.1}{60}$. Cross multiply: $60x = (5.1)(100)$. Simplify: $x = \frac{510}{60}$. Thus, $x = 8.5$, which means Toby's weight gain was 8.5%.

547. **d.** In this problem, you must first determine that, because 575 is the 10%-off price, it is also 90% of the original price. So, 575 is 90% of what number? First, change 90% to a decimal: 0.90. Then: $\frac{575}{0.90} = 575 ÷ 0.90 = 638.888$. Now round up to $638.89.

548. **a.** To solve this problem, change the percent to a decimal and multiply: 0.0525 × 380 = 19.95.

549. **b.** Let x equal the total number of minutes it will take Erin to complete her homework. Since it took her 42 minutes to complete 70% of her homework, set up the proportion $\frac{42}{x} = \frac{70}{100}$. Cross multiply to get $70x = 4,200$. Divide each side of the equation by 70 to get $x = 60$. Since it will take her 60 minutes to complete her homework and she has worked 42 minutes, it will take her 60 − 42 = 18 minutes to finish.

550. **d.** To determine the percent, first determine the original length of the rope: 8 + 7 + 5 = 20. Then set up an equation. Knowing that $x\% = \frac{x}{100}$, the equation is: $\frac{x}{100} = \frac{5}{20}$. Cross multiply: $20x = (5)(100)$. Simplify: $x = \frac{500}{20}$. Thus, $x = 25$, which means the shortest piece is 25% of the original rope length.

551. **c.** To find the whole when the percent is given, first set up an equation: 200 = 40% of the house payment. Convert the percent to a decimal: 200 = 0.4 × HP. Solve: 200 = 0.4 × 500.

552. **c.** Container B holds 12% more than container A. Convert the percent to a decimal and set up the equation; B = 0.12 × 8 + 8. Solve: B = 0.96 + 8; B = 8.96.

553. **a.** Simply set up the equation in the manner in which the problem is written. Since $x\% = \frac{x}{100}$, the equation is $\frac{x}{100} = \frac{.35}{1.40}$. Cross multiply: $1.40x = (0.35)(100)$. Simplify: $x = \frac{35}{1.40}$. Thus, $x = 25$, which means $0.35 is 25% of $1.40.

554. **d.** This is a two-step problem. First, you must determine the number of employees before the new hiring: 30 = 5% of the Original Work Force. Change the percent to a decimal: 30 = 0.05 × OWF. Solve: 30 = 0.05 × 600. Six hundred represents the number of employees before the new hiring. To find the total workforce now, simply add: 600 + 30 = 630.

555. **b.** This is a two-step problem involving multiplication and addition. First, determine how many cards were sold on Saturday: 0.05 × 200 = 10. That leaves 190 cards. Then, find out how many cards were sold on Sunday: 0.10 × 190 = 19. Next, add the cards that were sold: 10 + 19 = 29. Finally, subtract from the original number: 200 − 29 = 171.

556. **d.** To find the number when percent is given, change the percent to a decimal then set up the equation: $9.50 = 0.45 \times$ money in wallet. Solve: $9.5 = 0.45 \times x$. Therefore, $\frac{9.5}{0.45} = x$, so $x = \$21.11$.

557. **c.** This problem requires both multiplication and addition. First, to determine the amount of the raise, change the percent to a decimal and multiply: $0.0475 \times 27{,}400 = 1{,}301.5$. Then, add this amount to the original salary: $1{,}301.50 + 27{,}400 = 28{,}701.50$.

558. **c.** To find the number of days, multiply: $365 \times 0.15 = 54.75$. Round up to 55.

559. **d.** First, determine the percent of time that the station is NOT playing classical music. Subtract from 100%: $100 - 20 = 80$. Eighty percent of the time the station does NOT play classical music. Then change the percent to a decimal and multiply: $0.8 \times 24 = 19.2$.

560. **a.** First, change 70% to a decimal, which is 0.7. Then multiply: $70 \times 0.7 = 49$.

Set 36 (Page 82)

561. **b.** To find the percentage of people who said they rode at least three times a week, divide 105 by 150: $105 \div 150 = 0.7$, which is 70%. $0.7 \times 100{,}000 = 70{,}000$.

562. **d.** First, calculate the number of students that do not participate in extracurricular activities: $360 - 150 = 210$ students. Next, find the percent 210 is of 360 by setting up the proportion $\frac{210}{360} = \frac{x}{100}$. Cross multiply to get $360x = 21{,}000$. Divide each side of the equation by 360 to get $x \approx 58.33$, which rounds to 58%.

563. **b.** There are three steps involved in solving this problem. First, convert 4.5% to a decimal: 0.045. Multiply that by $26,000 to find out how much the salary increases. Finally, add the result ($1,170) to the original salary of $26,000 to find out the new salary, $27,170.

564. **c.** First you must subtract the percentage of the installation cost during construction (1.5%) from the percentage of the installation cost after construction (4%). To do this, begin by converting the percentages into decimals: $4\% = 0.04$; $1.5\% = 0.015$. Now subtract: $0.04 - 0.015 = 0.025$. This is the percentage of the total cost that the homeowner will save. Multiply this by the total cost of the home to find the dollar amount: $0.025 \times \$150{,}000 = \$3{,}750$.

565. **d.** Eighty out of 100 is 80%. Eighty percent of 30,000 is 24,000.

566. **b.** Since Lynn has read 157 out of 435 pages, set up the proportion $\frac{157}{435} = \frac{x}{100}$ to find the percent. Cross multiply to get $435x = 15{,}700$. Divide each side of the equation by 435 to get $x \approx 36.09$, which rounds to 36%.

567. **b.** Find 40% of 75 by multiplying 0.40×75, which is equal to 30.

568. **c.** First, find 75% of the pump's maximum rate which is $(0.75)(100) = 75$ gallons per minute. Next, find the time required using the equation *rate $\times$ time = amount pumped*; $75T = 15{,}000$; $T = \frac{15{,}000}{75}$; $T = 200$ minutes.

569. **a.** $\frac{1}{3} \times 0.60 = 0.20 = 20\%$

570. **a.** Twenty percent of 1,800 calories can be calories from fat; 20% of 1,800 equals $(0.2)(1{,}800)$, or 360 calories from fat are allowed. Since there are nine calories in each gram of fat, you should divide 360 by 9 to find that 40 grams of fat are allowed.

571. **b.** Use the formula beginning with the operation in parentheses: $98 - 32 = 66$. After that, multiply 66 by $\frac{5}{9}$, first multiplying 66 by 5 to get 330. 330 divided by $9 = 36.66$, which is rounded up to 36.7.

572. **c.** Add 9% + 15% to get 24%, since they are both calculated from the cost of the meal. Then find 24% of $24.95 by multiplying 0.24 × 24.95 = 5.988, which rounds to $5.99. Add $24.95 + $5.99 = $30.94.

573. **c.** Divide 1,132 by 4,528 to get 0.25. Change the decimal to a percent by multiplying by 100. The solution is 25%.

574. **a.** Add each of the known sectors, and subtract the sum from 100% to get 12%.

575. **d.** Since the rent sector is labeled 38%, find 38% of $2,450: 0.38 × 2,450 = $931.

576. **b.** Find 15% of $34 and subtract from the original price: 0.15 × 34 = 5.10; $34.00 − $5.10 = $28.90.

Set 37 (Page 84)

577. **c.** The difference between 220 and this person's age = 220 − 30, or 190. The maximum heart rate is 90% of 190, or (0.9)(190) = 171.

578. **a.** Find 20% of 30 and add it to the current class size of 30: 0.20 × 30 = 6; 30 + 6 = 36.

579. **a.** To find the percent of discount, divide the amount of savings by the original price and multiply by 100. The amount of savings is $350 − $279, which is equal to $71. Divide 71 by 350 to get 0.202. Multiply by 100 and round for a result of 20%.

580. **a.** The prescription is 50% effective for half (or 50%) of the symptoms, so it eliminates (0.50)(0.50) = 0.25, which is equal to 25% of all symptoms.

581. **b.** Since there are 171 men and 129 women, there are 171 + 129 = 300 people enrolled in the program. The percent of women can be found by dividing 129 by 300, which is equal to 0.43, and then multiplying by 100 to change the decimal to a percent: 0.43 × 100 = 43%.

582. **d.** The third and fourth quarters are 54% and 16%, respectively. This adds to 70%.

583. **c.** If 40% of those students are involved in the music program, then 100% − 40% = 60% are NOT involved in the music program. Find 60% of 315; 0.60 × 315 = 189.

584. **d.** Four inches is equal to 16 quarter inches, which is equal to (16)(2 feet) = 32 feet. You could also set up the problem as a proportion, so that $\frac{(\frac{1}{4})}{2\text{ feet}} = \frac{4}{x}$. Cross multiplying, you would get $(\frac{1}{4})x = 8$. Multiplying both sides by 4, you would get $x = 32$.

585. **d.** To find 150% of $45, change the percent to a decimal and multiply: 1.50 × 45 = 67.50. Since this is the markup of the price, add $67.50 to $45 to find the new price of $112.50.

586. **c.** You must break the 92,000 into the amounts mentioned in the policy: 92,000 = 20,000 + 40,000 + 32,000. The amount the policy will pay is (0.8)(20,000) + (0.6)(40,000) + (0.4)(32,000). This is 16,000 + 24,000 + 12,800, which equals $52,800.

587. **b.** First, find 50% of 100 gallons per minute which is (0.50)(100) = 50 gallons per minute. Next, convert the units of time from minutes to hours. 50 gallons per minute × 60 minutes = 3,000 gallons per hour. Now, use the formula *amount pumped = rate × time* and plug in the numbers. Amount pumped = (3,000)(6) = 18,000 gallons.

588. **b.** The number of papers graded is arrived at by multiplying the rate for each grader by the time spent by each grader. Melissa grades five papers per hour for three hours, or 15 papers; Joe grades four papers per hour for two hours, or eight papers, so together they grade 23 papers. Because there are 50 papers, the percentage graded is $\frac{23}{50}$, which is equal to 46%.

589. **c.** In one hour Dimitri can do 40% of 40 problems = $(0.40)(40) = 16$ problems; 16 problems is to 40 problems as 1 hour is to x hours: $\frac{16}{40} = \frac{1}{x}$. Cross multiply: $16x = 40$. Simplify: $x = \frac{40}{16} = 2.5$ hours.

590. **c.** First, find out how much the population will increase by multiplying $2,500 \times 0.03 = 75$. Then, add this amount to the current population to get the answer: $2,500 + 75 = 2,575$.

591. **a.** If the gas station is 57.8% of the way to the Chen's relative's house, it is $(0.578)(75) = 43.35$ miles from the Chen's house. To solve, subtract this distance from the total distance, $75 - 43.35 = 31.65$ miles. The gas station is about $31\frac{2}{3}$ miles from the Chen's relative's house.

592. **d.** Thirteen percent had not read books; therefore, 87% had; 87% is equal to 0.87; $0.87 \times 2,500 = 2,175$ people.

Set 38 (Page 87)

593. **d.** Change the percent to a decimal: $4\% = 0.04$. Now multiply: $500 \times 0.04 = 20$.

594. **a.** Multiply the percentages by one another ($30\% = 0.30$; $15\% = 0.15$.): $0.30 \times 0.15 = 0.045$ or 4.5%.

595. **d.** If 20% are eating the special, 80% are not. This means 40 people represent 80% of the number of people eating at the restaurant. So $0.80x = 40$ or $x = \frac{40}{0.80}$. Thus $x = 50$ people.

596. **d.** An average of 90% is needed of a total of 500 points: $500 \times 0.90 = 450$, so 450 points are needed. Add all the other test scores together: $95 + 85 + 88 + 84 = 352$. Now subtract that total from the total needed in order to see what score the student must make to reach 90%: $450 - 352 = 98$.

597. **d.** Change the fraction to a decimal by dividing the numerator of 3 by the denominator of 8; $3 \div 8 = 0.375$. Convert the decimal to a percent by multiplying by 100; $0.375 = 37.5\%$.

598. **c.** Find 90% of 90 by multiplying 0.90×90. The solution is 81.

599. **a.** The easiest way to solve this problem is to convert $5\frac{1}{2}\%$ to a decimal: 0.055. Then multiply: $200 \times 0.055 = 11$.

600. **c.** Set up the proportion $\frac{part}{whole} = \frac{\%}{100}$, and substitute 12 for the part and 20 as the percent. The proportion becomes $\frac{12}{x} = \frac{20}{100}$. Cross multiply to get $20x = 1,200$. Divide each side of the equation by 20 to get $x = 60$.

601. **a.** A ratio is given: 3 out of 12. Divide to find the decimal equivalent: 3 divided by $12 = 0.25$. Convert to percent form by multiplying by 1 in the form of $\frac{100}{100}$: $(0.25)(\frac{100}{100}) = \frac{25}{100}$ or 25%.

602. **b.** This is a multiplication problem involving a percent; 40% is equal to 0.4, so you must multiply the whole number by the decimal: $0.4 \times 8 = 3.2$.

603. **c.** Divide 18 by 45 to get 0.4. Change the decimal to a percent by multiplying by 100: $0.4 = 40\%$.

604. **d.** First, change 15% to a decimal and multiply: $0.15 \times \$26,000 = \$3,900$.

605. **c.** This percent problem asks that you find the whole when only a percent is known. Convert 325% to a decimal: $325\% = 3.25$. Now multiply the amount Patty usually spends by the percentage of increase: $\$4.75 \times 3.25 = \$15.44$. Since Patty spent $15.44 more than she usually spends, you must add the amount she usually spends to the amount of increase: $\$4.75 + \$15.44 = \$20.19$.

606. a. Subtract 196 from 230 to find the number of pounds he has lost: 230 − 196 = 34. Then divide 34 by the original weight of 230, and change the decimal to a percent: 34 ÷ 230 ≈ 0.148 ≈ 14.8%, or about 15%.

607. b. Calculate 5% of $45 by changing the percent to a decimal and multiplying: 0.05 × 45 = 2.25. Since the customer is getting a discount, solve $45.00 − $2.25 to get a final price of $42.75.

608. b. Convert 94% to a decimal to get 0.94, then multiply: 250 × 0.94 = 235.

Set 39 (Page 90)

609. c. Since there are eight equal sections shaded out of a total of 20, 8 ÷ 20 = 0.4 which is equal to 40%.

610. d. Subtract 28% from 100% to get 72%, which is the percent that represents Tim's check after taxes are taken out. Since $363.96 is equal to 72% of his gross pay, set up and solve the proportion $\frac{363.96}{x} = \frac{72}{100}$. Cross multiply to get 72x = 36,396. Divide each side of the equal sign by 72 to get x = $505.50.

611. b. In this two-step problem you must first determine what percent of the dresses are NOT designer dresses by subtracting; 100% − 8% = 92%. Now change 92% to a decimal and multiply: 0.92 × 300 = 276.

612. d. In this problem you need to find the whole when the percent is given. Since $48 is equal to 15% of the weekly budget, set up the proportion $\frac{48}{x} = \frac{15}{100}$. Cross multiply to get 15x = 4,800. Divide each side of the equation by 15 to get x = $320.

613. d. To find what percent one number is of another, first write out an equation. Since x% $= \frac{x}{100}$ the equation is $\frac{x}{100} = \frac{4}{26}$. Cross multiply

to get: 26x = (4)(100), or $x = \frac{400}{26}$, which makes x = 15.4. So approximately 15% of the pies are peach. (Check your answer: 26 × 0.15 = 3.9; round up to 4.)

614. d. The difference in the two weeks is $36 − $4 = $32. To find the percent of increase, compare the difference to the original amount by setting up the proportion $\frac{32}{4} = \frac{x}{100}$. Cross multiply to get 4x = 3,200. Divide each side of the equation by 4 to get 800%.

615. d. To find the amount of markup, find 25% of $34.99 by multiplying; 0.25 × 34.99 = $8.75. Then, the original price plus the markup equals $43.74.

616. d. Since Martha saved 35%, she paid $298.35, which was 65% of the original price. Set up the proportion $\frac{298.35}{x} = \frac{65}{100}$. Cross multiply to get 65x = 29,835. Divide each side of the equation by 65 to get x = $459.

617. c. To solve this problem, first multiply the cost of one doll by six, so Barbara buys six dolls originally valued at $1,800. Now change the percent to a decimal (3.5% = 0.035); $1,800 × 0.035 = $63, for the amount Barbara saves.

618. d. First, find out how many rooms Martin has already cleaned. Change the percent to a decimal: 40% = 0.4. (Remember zeros to the right of the digit that follows the decimal point do not change the value.) Now take 32 × 0.4 = 12.8, which you can round up to 13. Now, subtract: 32 rooms − 13 rooms already cleaned = 19 rooms left to clean. Alternatively, you know that if Martin has cleaned 40% of the 32 rooms, he still has 60% left to go: 32 × 0.6 = 19.2 or about 19 rooms.

619. b. The original length of the rope before it was cut is 5 + 4 + 3 = 12 inches. The longest piece is 5 inches. Five out of 12 becomes $\frac{5}{12} =$ $0.416 \approx 42\%$.

620. **c.** First, change the percent to a decimal and then multiply: $60\% = 0.60$. The problem becomes $0.60 \times 500 = \$300$.

621. **b.** Simply set up the equation in the manner in which the problem is written. Since $x\% = \frac{x}{100}$, the equation is $\frac{x}{100} = \frac{0.40}{1.30}$. Cross multiply: $1.30x = (0.40)(100)$. Simplify: $x = \frac{40}{1.30}$. Thus, $x = 30.7$, which means that $\$0.40$ is about 31% of $\$1.30$.

622. **a.** Since the ratio of males to females is three to two, for every three males there are two females. Therefore, there are two females out of every five people $(2 + 3 = 5)$. Change the ratio to a fraction and then divide to get a decimal; $\frac{2}{5} = 0.4$. Multiply by 100 to change to a percent: $0.4 \times 100 = 40\%$.

623. **b.** Since the ratio of rainy days to sunny days is three to four, for every three rainy days there are four sunny days. Therefore, there are four sunny days out of every seven days $(3 + 4)$. Change the ratio to a fraction and then divide to get a decimal: $\frac{4}{7} \approx 0.57$. Multiply by 100 to change to a percent: $0.57 \times 100 = 57\%$.

624. **d.** First, determine how many candies Toby ate for dessert: $0.25 \times 48 = 12$. That leaves $48 - 12 = 36$ candies for Toby to share with his friends. Next, determine how many candies Toby and his friends shared: $0.25 \times 36 = 9$. That leaves $36 - 9 = 27$ candies.

Set 40 (Page 92)

625. **c.** Hilga and Jerome's initial distance apart equals the sum of the distance each travels in 2.5 hours. Hilga travels a distance of $(2.5)(2.5) = 6.25$ miles, while Jerome travels $(4)(2.5) = 10$ miles. This means that they were $6.25 + 10 = 16.25$ miles apart.

626. **a.** Fifteen percent of 40 is $0.15 \times 40 = 6$. Six is the number who chose dry dog food. Therefore, $40 - 6 = 34$ is the number who chose canned dog food.

627. **a.** To figure what percentage 1 is of 8, the formula is $\frac{1}{8} = \frac{x}{100}$. Cross multiply: $100 = 8x$. Divide both sides by 8 to get $x = 12.5$.

628. **d.** To figure the total, the formula is $\frac{4}{x} = \frac{20}{100}$. Cross multiply: $4 \times 100 = 20x$ or $400 = 20x$, or $x = 20$.

629. **c.** To figure what percentage 15 is of 60 (minutes in an hour), the formula is $\frac{15}{60} = \frac{x}{100}$: $15 \times 100 = 60x$. Divide both sides by 60 to get $x = 25\%$.

630. **a.** $15 \times 100 = 75x$; $x = 20$

631. **c.** $80 \times 100 = 320x$; $x = 25$

632. **b.** $200 \times 78 = 100x$; $x = 156$

633. **c.** First, you must change the percent to a decimal: $85\% = 0.85$. Now find the amount at which the property is assessed: $\$185,000 \times 0.85 = \$157,250$. Next, divide to find the number of thousands: $\$157,250 \div 1,000 = 157.25$. Finally, find the tax: $\$24.85 \times 157.25 = \$3,907.66$.

634. **c.** First, convert the percents to decimals: $5\% = 0.05$; $2.5\% = 0.025$. Now find the Owens's *down payment* by multiplying: $\$100,000$ (the original cost) $\times 0.05 = \$5,000$. Next, find the *MIP* by multiplying: $\$95,000$ (the mortgage amount) $\times 0.025 = \$2,375$: Finally, add: $\$5,000$ (down payment) $+ \$2,375$ (MIP) $+ \$1,000$ (other closing costs) $= \$8,375$.

635. **d.** The seller's $\$103,000$ represents only 93% of the sale price $(100\% - 7\%)$. The broker's commission is NOT 7% of $\$103,000$, but rather 7% of the whole sale price. The question is $\$103,000$ is 93% of what figure. So, let $\frac{x}{100} = \frac{103,000}{93} = 110,752.68$, rounded to $\$110,753$.

636. **c.** First, change the percents to decimals. Next, find the total commission: $115,000 × 0.06 = $6,900. Finally, find salesperson Simon Hersch's cut, which is 55% of the total price (45% having gone to broker Bob King): $6,900 × 0.55 = $3,795.

637. **a.** First, find $\frac{1}{12}$ of the tax bill: $18,000 ÷ 12 = 1,500. Now find 3% of the gross receipts: $75,000 × 0.03 = 2,250. Now add: $1,000 (rent) + $1,500 (percent of annual tax bill) + $2,250 (percent of gross receipts) = $4,750.

638. **b.** First, find the amount of assessment: $325,000 × 0.90 = $292,500. Now find the number of thousands: $292,500 ÷ 1,000 = 292.5. Next, find the tax rate for a whole year: 292.5 × $2.90 = 848.25. Now find the tax rate for half of the year: $848.25 ÷ 2 = $424.13.

639. **a.** First, find the amount that the seller must pay out: $395,000 × 0.055 = $21,725 (brokerage fee), plus $395,000 × 0.04 = $15,800 (settlement fees), plus the $300,000 loan. Now subtract these from the gross proceeds of $395,000 − $21,725 − $15,800 − $300,000 = $57,475 net proceeds.

640. **a.** The property today is valued at 118% of purchase price: $585,000 × 1.18 = $690,300.

► Section 5—Algebra

Set 41 (Page 96)

641. **a.** When solving for x, isolate the variable by dividing both sides of the equation by 6. The correct answer is 7.

642. **c.** Ask the question, "What number divided by 72 equals 2?" The correct answer is 144.

643. **b.** Isolate the variable by subtracting 50 from both sides of the equation. Then divide both sides by 5. The correct answer is 20.

644. **a.** The square of a number is the number times itself. When you are looking at the answer choices, the only ones that would be possible are either choices **a** or **c**. Trial and error shows that $9 \times 9 = 81$. Then, multiply 81 by 9. The product is 729, so the correct answer is 9.

645. **d.** Start by adding 16 to 20. The sum is 36. Then, divide 36 by 6. The correct answer is 6.

646. **d.** One hundred percent of 100 is 100. Add it to 25% of 100, which is 25, to get 125. The correct answer is 100.

647. **c.** To find $\frac{1}{7}$ of 42, divide 42 by 7. The result is 6. Then, add 8: $6 + 8 = 14$.

648. **c.** First, find the product of the given values, which is 12. Then, square that number. Twelve squared equals 144. The correct answer is 144.

649. **a.** A prime number is a number that has exactly two factors: one and itself. The first four prime numbers are 2, 3, 5, and 7. The sum of these numbers is 17.

650. **d.** The correct answer is 1.

651. **c.** A common error is to think that 1 is a prime number. But, it is not, because one has only 1 factor: itself. Two is the correct answer.

652. **b.** Subtract both 6 and 9 from 49. The correct answer is 34.

653. **d.** Forty-two is approximately half of 80. As a percent, this is represented as 51%.

654. **b.** Divide 81 by 9 to find the complementary factor: $81 \div 9 = 9$. Then, multiply that number by 3: $9 \times 3 = 27$.

655. **c.** This problem can be represented by the following equation: $2(x \times 2x) = 196$. Solve for x: $2(2x^2) = 196$; $4x^2 = 196$; $x^2 = 49$; $x = 7$.

656. **d.** Subtract 5 from 25 to get 20. Then, multiply 20 by 5 to find the correct answer of 100.

Set 42 (Page 98)

657. **b.** Twelve-sixths of any number, x, is equivalent to twice the number, or $2x$. Solve for x: $135 - 75 = 2x$; $60 = 2x$; $x = 30$.

658. **a.** The property in the statement shows that the sum of the numbers is the same even if the order is changed. This is an example of the Commutative Property of Addition.

659. **d.** When the given values are plugged into the expression, it reads: $(-3)(6) - 6(-5)$; $-18 + 30 = 12$.

660. **c.** Subtract to find the correct answer, 64.

661. **d.** Let x equal the number sought. Begin the solution process by breaking up the problem into smaller parts: a number three times larger $= 3x$, 10 added to the number $= x + 10$. Combining terms we have: $3x = x + 10$. Simplify: $2x = 10$, or $x = 5$.

662. **d.** The symbol means that a is "less than or equal to" 5. The only choice that makes this statement false is 6.

663. **c.** First, combine like terms on the left side of the equation to get $2p - 10 = 16$. Add 10 to both sides of the equation: $2p = 26$. Divide both sides of the equation by 2: $p = 13$.

664. **d.** The equation is simply $50 + 3x = 74$; $3x = 24$; $x = 8$.

665. **a.** Let x equal the number sought. Four more than three times a number means $(3x + 4)$, so the expression becomes $(2)(3x + 4) = 20$. Simplify: $6x + 8 = 20$ or $6x = 12$. Thus, $x = 2$.

666. **c.** One of the most vital steps in solving for an unknown in any algebra problem is to isolate the unknown on one side of the equation.

667. **b.** Substitute 8 for y in the expression and perform the operations: $x = 4 + 6(8)$; $x = 52$.

668. **b.** Substitute the values of each letter and simplify. The expression becomes $\frac{(1)(3) + (3)(6)}{(1)(3)}$, which simplifies to $\frac{3 + 18}{3}$ after performing multiplication. Add $3 + 18$ in the numerator to get $\frac{21}{3}$, which simplifies to 7.

669. **c.** The Associative Property of Multiplication shows that even though the grouping of the numbers changes, the result is the same. Choice **c** changes the grouping of the numbers by placing different variables within the parentheses, while both sides of the equation remain equal.

670. **b.** To find the greatest common factor of $3x^2$, $12x$, and $6x^3$ first start with the coefficients, or numbers in front of the variables. The largest number that divides into 3, 6, and 12 without a remainder is 3. With the variables, the smallest exponent on x is 1, so x^1, or x, is the largest variable factor. Therefore, the greatest common factor of all three terms is $3x$.

671. **a.** Since the solution to the problem $x + 25 = 13$ is -12, choices **b**, **c**, and **d** are all too large to be correct.

672. **d.** The first step in solving this problem is to add the fractions to get the sum of $\frac{4x}{4}$. This fraction reduces to x.

Set 43 (Page 100)

673. **b.** Because the integers must be even, the equation $n + (n + 2) + (n + 4) = 30$ is used. This gives $3n + 6 = 30$; $3n = 24$; $n = 8$. Therefore, 8 is the first number in the series. Choice **a**, (9, 10, 11) would work, but the numbers aren't even integers.

674. **d.** The factors of a trinomial are often in the form $(x + a)(x + b)$ where a and b represent numbers whose sum is the coefficient in the middle term of the trinomial and whose product is the last term in the trinomial. In this case, the sum is 6 and the product is 9. The two numbers that satisfy this are 3 and 3. Therefore, express the factors as the two binomials by replacing a and b with 3 and 3: $(x + 3)(x + 3) = (x + 3)^2$.

675. **a.** Cross multiply: $(2x)(48) = (16)(12)$; $96x = 192$. Thus, $x = 2$.

676. **a.** To solve the inequality $12x - 1 < 35$, first solve the equation $12x - 1 = 35$. In this case, the solution is $x = 3$. Replace the equal sign with the *less than* symbol ($<$): $x < 3$. Since values of x *less than* 3 satisfy this inequality, 2 is the only answer choice that would make the inequality true.

677. **d.** $(x^2 + 4x + 4)$ factors into $(x + 2)(x + 2)$. Therefore, one of the $(x + 2)$ terms can be canceled with the denominator. This leaves $(x + 2)$.

678. **a.** The correct answer is $x(x^2 + 6)$.

679. **a.** Set up an equation and solve for the variable: $66 + x = 15$; $x = -51$.

680. **d.** Multiply using the distributive property, being sure to add the exponents of like bases; $4n^2(5np + p^2)$ becomes $4n^2 \times 5np + 4n^2 \times p^2$. This simplifies to $20n^3p + 4n^2p^2$.

681. **d.** Factor the numerator of the fraction to two binomials: $(x + 5)(x - 3)$. Cancel the common factors of $x + 5$ from the numerator and denominator. The result is $x - 3$.

682. **b.** Since $y \times y = y^2$, then $2y(y)$ is equal to $2y^2$.

683. **d.** x times x^2 is x^3; x times y is xy, so the solution to the problem is $3x^3 + xy$.

684. **d.** The key word *product* tells you to multiply. Therefore, multiply the coefficients of 2 and 3, and multiply the variables by adding the exponents of like bases. Keep in mind that $x = x^1$; $(3x^2y) \times (2xy^2)$ becomes $3 \times 2 \times x^2 \times x \times y \times y^2$. This simplifies to $6x^3y^3$.

685. **c.** This equation is a proportion, expressing the equivalence of two ratios.

686. **b.** Break the statement down into smaller parts. The first part, *the sum of two numbers,* a *and* b, can be translated $(a + b)$. This part needs to be in parentheses to ensure that the correct order of operations is executed. The second part, *divided by a third number,* c, takes the first part and divides it by c. This now becomes $(a + b) \div c$.

687. **d.** To solve this problem, you must first find the common denominator, which is 6. The equation then becomes $\frac{3x}{6} + \frac{x}{6} = 4$; then, $\frac{4x}{6} = 4$; and then $4x = 24$, so $x = 6$.

688. **c.** Since 45 is nine times larger than five in the denominators, convert to an equivalent fraction by multiplying 2 by 9 to get the resulting numerator: $2 \times 9 = 18$, so $x = 18$. Another way to solve this problem is to cross multiply to get $5x = 90$. Divide both sides of the equation by 5 to get $x = 18$.

Set 44 (Page 102)

689. **d.** Solve for x: $\frac{1}{5}x - 6 = 3$; $\frac{1}{5}x = 9$; $x = 45$.

690. **c.** Solve for y: $8y - 3 = 1$; $8y = 4$; $y = \frac{1}{2}$.

691. **d.** Substitute 5 for x in the equation, so that there is only one variable for which to solve: $y = 2(5^2 + 10) - 3$; $y = 2(35) - 3$; $y = 70 - 3$; $y = 67$.

692. **c.** In order to find an equivalent fraction, you need to perform the same action on both the numerator and the denominator. One way to solve for x is to ask the question, "What is multiplied by 19 (the denominator) to get a product of 76?" Divide: $76 \div 19 = 4$. Then, multiply the numerator by 4 in order to find the value of x: $4 \times 1 = 4$.

693. **a.** The correct answer is 5.

694. **d.** All are possible solutions to the inequality; however, the largest solution will occur when $\frac{1}{3}x - 3 = 5$. Therefore, $\frac{1}{3}x = 8$; $x = 24$.

695. **c.** A linear equation in the form $y = mx + b$ has a slope of m and a y-intercept of b. In this case you are looking for an equation where $m = -2$ and $b = 3$. The equation is $y = -2x + 3$.

696. **b.** The simplest way to solve this problem is to cancel the a term that occurs in both the numerator and denominator. This leaves $\frac{(b - c)}{bc}$. This is $\frac{(4 - (-2))}{4(-2)}$, which simplifies to $-\frac{3}{4}$.

697. **c.** First, find the least common denominator and multiply each term on both sides of the equation by this number. The least common denominator in this case is 10. When 10 is multiplied by each term, the equation simplifies to $2x - x = 40$, which becomes $x = 40$.

698. **c.** Slope is equal to the change in y values divided by the change in x values. Therefore, $\frac{(3-(-1))}{(2-0)} = \frac{4}{2} = 2$. The intercept is found by putting 0 in for x in the equation $y = 2x + b$: $-1 = 2(0) + b$; $b = -1$. Therefore, the equation is $y = 2x - 1$.

699. **a.** Let x equal the number sought. Working in reverse order we have: The sum of that number and six becomes $(x + 6)$, the product of three and a number becomes $3x$. Combine terms: $(x + 6) - 3x = 0$. Simplify: $2x = 6$ or $x = 3$.

700. **a.** To find the slope of a linear equation, write the equation in the form $y = mx + b$ where m represents the slope and b represents the y-intercept. To change $3y - x = 9$ to $y = mx + b$ form, first add x to both sides of the equation: $3y - x + x = x + 9$: $3y = x + 9$. Then divide both sides by 3: $\frac{3y}{3} = \frac{x}{3} + \frac{9}{3}$, which simplifies to $y = \frac{1}{3}x + 3$. (Remember that $\frac{x}{3} = \frac{1}{3}x$.) The number in front of x is $\frac{1}{3}$, and this represents m, or the slope of the line.

701. **b.** Two equations are used: $T = 4O$, and $T + O = 10$. This gives $5O = 10$, and $O = 2$. Therefore, $T = 8$. The number is 82.

702. **a.** The sum of three numbers means $(a + b + c)$, the sum of their reciprocals means $(\frac{1}{a} + \frac{1}{b} + \frac{1}{c})$. Combine terms: $(a + b + c)(\frac{1}{a} + \frac{1}{b} + \frac{1}{c})$. Thus, choice **a** is the correct answer.

703. **a.** Only like terms can be added: $4x - 7y + 7x + 7y$; $4x + 7x$ and $-7y + 7y$. The y terms cancel each other out, leaving $11x$ as the correct answer.

704. **b.** Let x equal the number sought. Begin by breaking the problem into parts: *88 is the result* becomes $= 88$. *One-half of the sum of 24 and a number* becomes $0.5(24 + x)$. *Three times a number* becomes $3x$. Combine terms: $3x - 0.5(24 + x) = 88$. Simplify: $3x - 12 - 0.5x = 88$, which reduces to $2.5x = 100$. Thus, $x = 40$.

Set 45 (Page 104)

705. **b.** A straight line that has a positive slope slants up to the right. Choice **a** has a negative slope and slants up to the left. Choice **c** is a horizontal line and has a slope of zero. Choice **d** is a vertical line and has no slope or an undefined slope.

706. **d.** Let G stand for the generic oatmeal and N for the name brand: $G = \frac{2}{3}N$; $G = \$1.50$; $\$1.50 \div \frac{2}{3} = \$2.25$.

707. **b.** You need to know D, the difference in total revenue. Let A equal total revenue at \$6 each, and B = total revenue at \$8 each. Therefore, D = A − B. You are given $A = 6 \times 400 = 2{,}400$ and $B = 8 \times 250 = 2{,}000$. Substitute: $D = 2{,}400 - 2{,}000$. Thus, D = 400 dollars.

708. **c.** You are asked to find B, the dollar amount of tips Betty is to receive. You know from the information given that $T + J + B = 48$, $T = 3J$, and $B = 4J$. The latter can be rearranged into the more useful: $J = B \div 4$. Substitute: $3J + J + B = 48$. This becomes $3(B \div 4) + (B \div 4) + B = 48$. Simplify: $(3 + 1)(B \div 4) + B = 48$, which further simplifies to $4(B \div 4) + B = 48$, or $2B = 48$. Thus, B = \$24.

709. **b.** Since the lines intersect at one point, there is one solution to this system of equations. They cross at the point (1,3).

710. **c.** The total bill is the sum of the three people's meals. You are solving for the father's meal. The mother's meal is $\frac{5}{4}F$, and the child's meal is $\frac{3}{4}F$. Therefore, $F + \frac{5}{4}F + \frac{3}{4}F = 48$. This simplifies to $3F = 48$; $F = \$16$.

711. **b.** You are solving for x, the cost per pound of the final mixture. You know that the total cost of the final mixture equals the total cost of the dried fruit plus the total cost of the nuts, or $C = F + N$. In terms of x, C is the cost per pound times the number of pounds in the mixture, or

$C = 7.5x$. Substitute: $7.5x = 3(6) + 7(1.5)$. Simplify: $7.5x = 18 + 10.5$ or $x = \frac{28.5}{7.5}$. Thus, $x = \$3.80$ per pound.

712. d. You want to know R = helicopter's speed in miles per hour. To solve this problem, recall that *rate* × *time* = *distance*. It is given that $T = 6{:}17 - 6{:}02 = 15$ minutes $= 0.25$ hour and $D = 20$ miles. Substitute: $r \times 0.25 = 20$. Simplify: $r = 20 \div 0.25$. Thus, $r = 80$ miles per hour.

713. c. $12 \times 5\% + 4 \times 4\% = x$ times 16; $x = 4.75\%$.

714. b. Let x equal D'Andre's rate. D'Andre's rate multiplied by his travel time equals the distance he travels; this equals Adam's rate multiplied by his travel time: $2x = D = 3(x - 5)$. Therefore, $2x = 3x - 15$, or $x = 15$ miles per hour.

715. d. The equation to describe this situation is $\frac{10 \text{ fish}}{\text{hour}}(2 \text{ hours}) = \frac{2.5 \text{ fish}}{\text{hour}}t$; $20 = 2.5t$; $t = 8$ hours.

716. b. Solve this problem by finding the weight of each portion. The sum of the weights of the initial corn is equal to the weight of the final mixture. Therefore, $(20 \text{ bushels})\frac{56 \text{ pounds}}{\text{bushel}} + (x \text{ bushels})\frac{50 \text{ pounds}}{\text{bushel}} = [(20 + x) \text{ bushels}]\frac{54 \text{ pounds}}{\text{bushel}}$. Thus, $20 \times 56 + 50x = (x + 20) \times 54$.

717. d. The total sales equal the sum of Linda and Jared's sales, or $L + J = 36$. Since Linda sold three less than twice Jared's total, $L = 2J - 3$. The equation $(2J - 3) + J = 36$ models this situation. This gives $3J = 39$; $J = 13$.

718. d. The amount done in one day must be found for each person. John strings $\frac{1}{4}$ of a fence in a day, and Mary strings $\frac{1}{3}$ of a fence in a day: $\frac{1}{4} + \frac{1}{3} = \frac{1}{x}$; multiplying both sides by $12x$ yields $3x + 4x = 12$; $x = \frac{12}{7} = 1\frac{5}{7}$ days.

719. b. One and a half cups equals $\frac{3}{2}$ cups. The ratio is six people to four people, which is equal to the ratio of $\frac{3}{2}$. By cross multiplying, you get $6(\frac{3}{2})$ equals $4x$, or 9 equals $4x$. Dividing both sides by 4, you get $\frac{9}{4}$, or $2\frac{1}{4}$ cups.

720. c. Since odd integers, such as 3, 5 and 7, are two numbers apart, add 2 to the expression: $x - 1 + 2$ simplifies to $x + 1$.

Set 46 (Page 107)

721. c. Multiply the two binomials using the distributive property so that each term from the first set of parentheses gets multiplied by each term of the second set of parentheses: $(x - 3)(x + 7) = x \cdot x + x \cdot 7 - 3 \cdot x - 3 \cdot 7$. Simplify the multiplication next: $x^2 + 7x - 3x - 21$. Combine like terms: $x^2 + 4x - 21$.

722. c. *Karl is four times as old as Pam* means $K = 4P$, and *Pam is one-third as old as Jackie* means $P = \frac{1}{3}J$. You are given $J = 18$. Working backward, you have: $P = \frac{1}{3}(18) = 6$; $K = 4(6) = 24$. The sum of their ages $= K + P + J = 24 + 6 + 18 = 48$.

723. c. In order to solve for x, get x alone on one side of the equation. First, add r to both sides of the equation: $s + r = 2x - r + r$. The equation becomes $s + r = 2x$. Then divide each side of the equation by 2: $\frac{s + r}{2} = \frac{2x}{2}$. Cancel the 2's on the right side of the equation to get a result of $\frac{s + r}{2} = x$, which is equivalent to answer choice **c**.

724. b. To solve this problem, set up the proportion 3 is to 25 as x is to 40: $\frac{3}{25} = \frac{x}{40}$. Cross multiply: $(3)(40) = 25(x)$. Solving for x gives 4.8, but since coolers must be whole numbers, this number is rounded up to 5.

725. b. A ratio is set up: $\frac{1.5}{14} = \frac{2.25}{x}$. Solving for x gives 21 pancakes.

726. a. Two equations are used: $A + B + C = 25$, and $A = C = 2B$. This gives $5B = 25$, and $B = 5$.

727. **c.** The problem is to find x, the number of gallons of 75% antifreeze. You know the final mixture is equal to the sum of the two solutions or M = A + B. In terms of x, M = 0.50(x + 4), A = 0.75x, and B = 0.30(4). Substitute: 0.50(x + 4) = 0.75x + 1.20. Simplify: 0.50x + 2 = 0.75x + 1.2, which reduces to 0.25x = 0.80. Thus, x = 3.2 gallons.

728. **a.** To move from one point to another on the line, you need to move up three units and over one to the right. Therefore, the line has a slope of 3. The line crosses the y-axis at +1 so the y-intercept is 1. Using the slope-intercept form of the equation ($y = mx + b$, where m = the slope and b = the y-intercept), the equation becomes $y = 3x + 1$.

729. **a.** The problem is to find x the number of pounds of raspberries. You know that total fruit purchased equals the sum of bananas and raspberries, or F = B + R. In terms of x, F = 2(x + 7), B = 7(0.50), and R = 4x. Substitute: 2(x + 7) = 3.5 + 4x. Simplify: 2x + 14 = 3.5 + 4x, which reduces to 2x = 10.5. Thus, x = 5.25 pounds.

730. **b.** The problem is to find x, the number of rental months needed to make the costs equal. This occurs when purchase cost equals rental cost, or P = R. You are given P = 400 and R = 50 + 25x. Substitute: 400 = 50 + 25x, which reduces to 25x = 350. Thus, x = 14 months.

731. **b.** Two equations are used: E = M + 10 and 20E + 25M = 1,460. This gives 20M + 200 + 25M = 1,460; 45M = 1,260; M = 28.

732. **c.** Use the FOIL method to find the answer. This stands for the order by which the terms are multiplied: First + Outside + Inside + Last: (3x × x) + (3x × –6) + (4x) + (4 × –6); 3x^2 – 18x + 4x – 24. The correct answer is 3x^2 – 14x – 24.

733. **c.** The unknown, I, equals the amount of interest earned. To solve this problem, it is necessary to use the formula given: *Interest = principal* × *rate* × *time* (I = *PRT*). First, change the percent to a decimal: $7\frac{1}{8}$ = 0.07125. Next note that nine months = $\frac{9}{12}$ or $\frac{3}{4}$ or 75% (0.75) of a year. Substitute: I = ($767)(0.07125)(0.75). Thus, I = $40.99.

734. **a.** You are asked to find F, Fluffy's age. Begin the solution by breaking the problem into parts: *Fluffy is half the age of Muffy* becomes F = $\frac{1}{2}$M, *Muffy is one-third as old as Spot* becomes M = $\frac{1}{3}$S, and *Spot is half the neighbor's age* becomes S = $\frac{1}{2}$N. You know the neighbor's age is 24 or N = 24. Substitute and work backward through the problem: S = $\frac{1}{2}$(24) = 12, M = $\frac{1}{3}$(12) = 4, F = $\frac{1}{2}$(4). Thus, F = 2. Fluffy is two years old.

735. **b.** Let x equal the unknown quantity of each denomination. You know that all the coins total $8.20 and that each denomination is multiplied by the same number, x. Therefore, 0.25x + 0.10x + 0.05x + 0.01x = 8.20. This reduces to (0.25 + 0.10 + 0.05 + 0.01)x = 8.20, or 0.41x = 8.20. Thus, x = 20 coins in each denomination.

736. **d.** The order of operations dictates addressing the exponents first. Then perform the operations within the parentheses. Finally, perform the operation outside the parentheses: Raising exponents by exponents is done by multiplying: 4 × 2 = 8; 3(36x^8); the correct answer is 108x^8.

Set 47 (Page 109)

737. **b.** Let x equal the amount of interest Samantha earns in one year. Substitute: I = (385) (.0485)(1). Thus, I = $18.67.

738. **a.** Let R equal Veronica's average speed. Recall that for uniform motion *Distance = rate × time* or $D = RT$. Substitute: $220 = R(5)$ or $R = \frac{220}{5}$. Thus, $R = 44$ miles per hour.

739. **b.** This is a special expression called a *perfect square*. The x terms cancel each other out, leaving just two terms in the expression; therefore, $(x + 5)(x - 5)$ is the correct factoring of the expression.

740. **c.** Each of the choices has the same numbers. The key to finding the correct answer to this item is choosing the correct signs. Look at the x value of the expression: $-2x$. That indicates that when the x terms are added, the result is -2. Choice **c** shows this product when multiplied out: $(x - 8)(x + 6)$.

741. **b.** Each of the answer choices has the same slope and y-intercept, so you need to look at the symbol used in the inequality. Since the line drawn is dashed and shaded above the line, the symbol used must be the greater than symbol, or $>$. This is used in choice **b**.

742. **d.** By adding the two equations vertically, you end up with $2y = 8$, so y must equal 4. Substitute 4 in for y in either original equation to get $x = 6$. Therefore, the point of intersection where the two lines are equal is $(6,4)$.

743. **b.** $150x = (100)(1)$, where x is the part of a mile a jogger has to go to burn the calories a walker burns in one mile. If you divide both sides of this equation by 150, you get $x = \frac{100}{150} = \frac{[2(50)]}{[3(50)]}$. Canceling the 50s, you get $\frac{2}{3}$. This means that a jogger has to jog only $\frac{2}{3}$ of a mile to burn the same number of calories a walker burns in a mile of brisk walking.

744. **b.** 5% of 1 liter $= (0.05)(1) = (0.02)x$, where x is the total amount of water in the resulting

2% solution. Solving for x, you get 2.5. Subtracting the 1 liter of water already present in the 5% solution, you will find that 1.5 (2.5 − 1) liters need to be added.

745. **c.** The problem is to find J, Joan's present age, in years. Begin by breaking the problem up into smaller parts: Joan will be twice Tom's age in three years becomes J + 3 = 2T; Tom will be 40 becomes T = 40. Substitute: J +3 = 2(40). Simplify: J = 80 −3, or J = 77 years old.

746. **d.** Cancel out the common factors of 3, x, and y between numerators and denominators, and then multiply across. The result is $\frac{xy^2}{12}$.

747. **d.** 3W equals water coming in, W equals water going out; 3W minus W equals 11,400, which implies that W is equal to 5,700 and 3W is equal to 17,100.

748. **a.** Set up an equation using one variable: Jim = 2 × Sally; Bill = 2 × Sally; $s + 2s + 2s = 60$; $5s = 60$; $s = 12$. So, Sally is 12 years old. The problem asks to find Jim's age; therefore, Jim is 24 years old.

749. **d.** You are asked to find x, minutes until the boat disappears. Recall that *Distance = rate × time* or $D = RT$ or $T = \frac{D}{R}$. We are given $D = 0.5$ mile and $R = 20$ miles per hour. Substitute: $T = \frac{0.5}{20} = 0.025$ hour. Convert the units of time by establishing the ratio 0.025 hours is to 1 hour as x minutes is to 60 minutes or $\frac{0.025}{1} = \frac{x}{60}$. Cross multiply: $(0.025)(60) = (1)(x)$, which simplifies to $x = 1.5$ minutes.

750. **a.** First, find out how long the entire hike can be, based on the rate at which the hikers are using their supplies; $\frac{\frac{2}{5}}{3} = \frac{1}{x}$, where 1 is the total amount of supplies and x is the number of days for the whole hike. Cross multiplying, you get $\frac{2x}{5} = 3$, so that $x = \frac{(3)(5)}{2}$, or $7\frac{1}{2}$ days for the length of the entire hike. This means that the hikers could go forward for 3.75 days altogether before they would have to turn around. They have already hiked for three days; 3.75 minus 3 equals 0.75 for the amount of time they can now go forward before having to turn around.

751. **d.** Factor the left side of the equation and set each factor equal to zero: $x^2 - 25 = (x - 5)$ $(x + 5)$; $x - 5 = 0$ or $x + 5 = 0$. Therefore, $x = 5$ or -5.

752. **d.** If Jason's price is $\frac{3}{4}$ of Lisa's, that would mean that if $63 is divided by 3, the quotient will be $\frac{1}{4}$. Add this to Jason's price, and the sum is Lisa's price: $63 \div 3 = 21; $63 + $21 = 84. Lisa's salon charges $84.

Set 48 (Page 111)

753. **b.** Take the reciprocal of the fraction being divided by, change the operation to multiplication, and cancel common factors between the numerators and denominators: $\frac{6a^2b}{2c} \div \frac{ab^2}{4c^4}$ becomes $\frac{6a^2b}{2c} \times \frac{4c^4}{ab^2} = \frac{12ac^3}{b}$.

754. **c.** Let x equal the number of oranges left in the basket. Three more than seven times as many oranges as five is $7(5) + 3 = 38$. Removing five leaves $x = 38 - 5 = 33$ oranges.

755. **b.** The problem is to find x, the number of ounces of candy costing $1 (or 100 cents) per ounce. The total cost of the mixture equals the sum of the cost for each type of candy, or $M = A + B$, where $A = 100x$, $B = 70(6)$, and $M = 80(6 + x)$. Substitute: $80(6 + x) = 100x + 420$. Simplify: $480 + 80x = 100x + 420$, which becomes $480 - 420 = 100x - 80x$, or $20x = 60$. Thus, $x = 3$ ounces.

756. **a.** Let x equal the number of hours to wash and wax the car if both work together. In one hour Dave can do $\frac{1}{4}$ of the job, while Mark can do $\frac{1}{3}$ of the job. In terms of x this means: $\frac{1}{4}x + \frac{1}{3}x = 1$ (where 1 represents 100% of the job). Simplify: $(\frac{1}{4} + \frac{1}{3})x = 1$ or $(\frac{3}{12} + \frac{4}{12})x = 1$. Thus, $\frac{7}{12}x = 1$ or $x = \frac{12}{7} = 1.7$ hours.

757. **b.** Let T equal the time it takes Sheri to walk to the store. Since she walks at a uniform rate, you can use the formula *Distance = rate × time* or $D = RT$. Substitute: $5 = 3T$ or $T = \frac{5}{3}$. Thus, $T = 1.67$ hours.

758. **a.** Let D equal the time Dee arrived before class. Choosing to represent time before class as a negative number, you have: Jeff arrived 10 minutes early means J = −10, Dee came in four minutes after Mae means D = M + 4, Mae, who was half as early as Jeff means M = $\frac{1}{2}$J. Substitute: M = −5, so D = −5 + 4 = −1. Thus, D = 1 minute before class time.

759. **d.** Substitute: $5 = (45)R(\frac{1}{12})$ or $5 = (45)R$ (0.0833). Thus, R = 1.33 = 133%.

760. **b.** Let x equal the number of people remaining in the room. You have: $x = 12 - (\frac{2}{3}(12) + 3)$ or $x = 12 - (8 + 3) = 12 - 11$. Thus, $x = 1$ person.

761. **b.** Let x equal the height of the ceiling. After converting 24 inches = 2 feet, you have $2 = 0.20x$ or $x = 10$ feet.

762. **d.** Let x equal the number of hours it takes Belinda to complete the job. In one hour the neighbor can do $\frac{1}{38}$ of the job, while Belinda can do $\frac{1}{x}$ of the job. Working together, they take 22 hours to complete 100% of the job or: $\frac{1}{38}(22) + \frac{1}{x}(22) = 1$ (where 1 represents 100% of the job). Simplify: $\frac{22}{38} + \frac{22}{x} = 1$ or $\frac{22}{x} = 1 - \frac{22}{38}$, which reduces to $\frac{22}{x} = \frac{16}{38}$. Cross multiply: $16x = (22)(38)$, or $x = 52.25$ hours.

763. **c.** Let x equal the number sought. If there are seven times as many candles as nine, there must be $x = 7 \times 9 = 63$ candles.

764. **a.** Take the square root of each of the values under the radical: $\sqrt{81} = 9$; $\sqrt{x^2} = x$; $\sqrt{y^2} = y$. Therefore, the radical becomes $\sqrt{\frac{81x^2}{y^2}} = \frac{9x}{y}$.

765. **d.** Equations in the form $y = mx + b$ have a slope of m. If each answer choice is transformed to this form, choice **d** has a negative slope: $6y + x = 7$ becomes $y = \frac{-1}{6}x + \frac{7}{6}$.

766. **a.** This is the same as the equation provided; each score is divided by three.

767. **a.** To simplify the radical, first find the square root of 64, which is 8. Then divide each exponent on the variables by 2 to find the square root of the variables. If the exponent is odd, the remainder stays inside the radical: $\sqrt{x^5} = x^2\sqrt{x}$ and $\sqrt{y^8} = y^4$. Thus, the result is $8x^2y^4\sqrt{x}$.

768. **c.** You are trying to find T, the number of minutes it will take the tire to completely deflate. The formula to use is *Pressure = rate × time*, or $P = RT$. In terms of T this becomes $T = \frac{P}{R}$. You are given $P = 36$ psi, $R = 3\frac{\text{psi}}{\text{per minute}}$, and therefore, $T = \frac{36}{3}$. Thus, $T = 12$ minutes.

Set 49 (Page 113)

769. **d.** Let x equal the number of months needed to pay off the loan. First, convert $\frac{5}{6}$ of a year to x months by establishing the ratio: five is to six as x is to 12, or $\frac{5}{6} = \frac{x}{12}$. Cross multiply: $(5)(12) = 6x$ or $x = 10$ months. Counting 10 months forward, you arrive at the answer, the end of December.

770. **d.** Square each side of the equal sign to eliminate the radical: $(\sqrt{b - 4})^2 = 5^2$ becomes $b - 4 = 25$. Add 4 to both sides of the equation: $b = 29$.

771. **c.** Since there is a common denominator, add the numerators and keep the denominator: $\frac{7w}{z}$.

772. **b.** You are trying to find x, the number of birds originally in the oak tree. Ten more birds landed means there are now $x + 10$ birds, a total of four times as many birds means the oak tree now has $4(x + 10)$ birds. In the maple tree, 16 less than 12 times as many birds as the oak tree had to begin with means there are $12x - 16$ birds in it. Set the two equations equal: $4(x + 10) = 12x - 16$. Simplify: $4x + 40 = 12x - 16$ or $8x = 56$. Thus, $x = 7$ birds.

773. **b.** You are looking for D, the number of dimes in the jar. Twice as many pennies as dimes means there are P = 2D number of pennies. The total dollar amount of coins in the jar is T = 0.25Q + 0.10D + 0.05N + 0.01P. Substitute: 4.58 = 0.25(13) + 0.10(D) + 0.05(5) + 0.01(2D). Simplify: 4.58 = 3.25 + 0.10D + 0.25 + 0.02D, which reduces to 0.12D = 1.08. Thus, D = 9 dimes.

774. **d.** Substitute -2 for x in the function and evaluate: $5(-2)^2 - 3 = 5(4) - 3 = 20 - 3 = 17$.

775. **b.** You are looking for x, the hiker's total trip time in hours, which will be twice the time it takes to get from his car to the lake. You are told that he travels two hours over smooth terrain. The time to walk the rocky trail is found by using the distance formula (*Distance = rate × time*) and rearranging to solve: $T = \frac{D}{R}$. Substitute: $T = \frac{5}{2} = 2.5$ hours. Therefore, $x = 2(2 + 2.5) = 9$ hours.

776. **d.** Total pressure is equal to P = O + N + A. You are given: N = 4, O = $\frac{1}{2}$N = 2, and A = $\frac{1}{3}$O = $\frac{2}{3}$; so P = 2 + 4 + $\frac{2}{3}$; P = $6\frac{2}{3}$ psi.

777. **b.** You are seeking D, the number of feet away from the microwave where the amount of radiation is $\frac{1}{16}$ the initial amount. You are given: radiation varies inversely as the square of the distance, or $R = 1 \div D^2$. When D = 1, R = 1, so you are looking for D when $R = \frac{1}{16}$. Substitute: $\frac{1}{16} = 1 \div D^2$. Cross multiply: $(1)(D^2) = (1)(16)$. Simplify: $D^2 = 16$, or D = 4 feet.

778. **a.** Choice **a** is a vertical line, where the change in x is zero. Vertical lines have an undefined slope.

779. **d.** First, add 4 to both sides of the equation: $\sqrt{2a + 6} - 4 + 4 = 6 + 4$. The equation simplifies to $\sqrt{2a + 6} = 10$. Square each side to eliminate the radical sign: $(\sqrt{2a + 6})^2 = 10^2$. The equation becomes $2a + 6 = 100$. Subtract 6 from each side of the equal sign and simplify: $2a + 6 - 6 = 100 - 6$; $2a = 94$. Divide each side by 2: $\frac{2a}{2} = \frac{94}{2}$. Therefore, $a = 47$.

780. **d.** Let F equal the final amount of money Willie will receive, which is his salary less the placement fee or F = S − P. You are given S = $28,000 and $P = \frac{1}{12}S$. Substitute: F = 28,000 − $\frac{1}{12}(28,000)$ or F = 28,000 − 2,333. Thus, F = $25,667.

781. **d.** The problem asks you to add $\frac{1}{3}$ of 60 and $\frac{2}{5}$ of 60. Let x equal the number sought. You have $x = \frac{1}{3}(60) + \frac{2}{5}(60)$, or $x = 20 + 24$. Thus, $x = 44$.

782. **a.** The 90% discount is over all three items; therefore, the total price is $(a + b + c) \times 0.9$. The average is the total price divided by the number of computers: $\frac{0.9 \times (a + b + c)}{3}$.

783. **b.** Let C equal the number of cherries. It is given that three apples and six oranges equals $\frac{1}{2}C$ or $9 = \frac{1}{2}C$. Therefore, C = 2(9) = 18.

784. **b.** Let L equal the number of gallons of gas lost, which is equal to the rate of loss times the time over which it occurs, or L = RT. Substitute: $L = (7)(\frac{1}{3}) = 2\frac{1}{3}$ gallons. Notice that the 14-gallon tank size is irrelevant information in this problem.

Set 50 (Page 117)

785. **c.** The problem can be restated as five hours is to 24 hours as x% is to 100%. This is the same as $\frac{5}{24} = \frac{x}{100}$.

786. **c.** Let x equal the number of pounds of white flour. The problem can be restated more usefully as five parts is to six parts as x pounds is to 48 pounds or $\frac{5}{6} = \frac{x}{48}$. Cross multiply: $(5)(48) = 6x$, or $x = \frac{240}{6}$. Thus, $x = 40$.

787. **a.** Let x equal the extra amount Timmy will earn by charging 25 cents instead of 10 cents per glass, which must be the difference in his total sales at each price. Therefore, x = sales at 0.25 per glass − sales at 0.10 per glass. Substitute: $x = 0.25(7) − 0.10(20) = 1.75 − 2.00$. Thus, $x = −0.25$. Timmy will lose money if he raises his price to 25 cents per glass.

788. **b.** Let J equal the number of miles away from school that Joe lives. You are given B = 5 and $T = \frac{1}{2}B = 2.5$. The distance between Bill and Tammy's house is (5 − 2.5). Since Joe lives half way between them, you have: $J = T + \frac{1}{2}(5 − 2.5)$. Substitute: $J = 2.5 + \frac{1}{2}(2.5)$. Thus, J = 3.75 miles.

789. **b.** Let V equal Valerie's annual salary. You know Pamela's earnings are six times Adrienne's, or P = 6A; Adrienne earns five times more than Beverly, or A = 5B; and Beverly earns $4,000, or B = 4,000. Working backward, you have A = 5(4,000) = 20,000, and P = 6(20,000) = 120,000. Finally, you are told that Valerie earns $\frac{1}{2}$ as much as Pamela, or $V = \frac{1}{2}P$, so $V = \frac{1}{2}(120,000)$. Thus, V = $60,000.

790. **b.** Let G equal Gertrude's age. One-fourth Gertrude's age taken away from Yolanda's becomes $Y - \frac{1}{4}G$, and twice Gertrude's age becomes 2G. Combine terms: $Y - \frac{1}{4}G = 2G$, which simplifies to $Y = (2 + \frac{1}{4})G$, or $Y = 2.25G$. Substitute: $9 = 2.25G$ or $G = 4$. Gertrude is 4 years old.

791. **a.** Let x equal the number of pounds of chocolate to be mixed. You know the mixture's total cost is the cost of the chocolates plus the cost of the caramels, or $M = A + B$. In terms of x, $M = 3.95(x + 3)$, $A = 5.95x$, while $B = 2.95(3)$. Combine terms: $3.95(x + 3) = 5.95x + 2.95(3)$. Simplify: $3.95x + 11.85 = 5.95x + 8.85$, or $11.85 - 8.85 = (5.95 - 3.95)x$, which becomes $2x = 3$. Thus, $x = 1.5$ pounds.

792. **b.** You want to find S, the rate Sharon charges to mow a lawn in dollars per hour. You are given Kathy's rate, which is $K = 7.50$, and you are told that $S = 1.5K$. Substitute: $S = 1.5(7.50)$. Thus, $S = \$11.25$ per hour.

793. **c.** The problem is to find H, the number of years Hazel will take to save $1,000. You are told Laura saves three times faster than Hazel, a ratio of 3:1. Therefore, $3L = H$. You are given $L = 1.5$ years. Substitute: $3(1.5) = H$, or $H = 4.5$ years.

794. **a.** First, get the common denominator of $6a^3$ by multiplying the first fraction by $3a^2$ in the numerator and denominator. The problem becomes $\frac{27a^2}{6a^3} - \frac{3w}{6a^3}$. Combine over the same denominator and cancel any common factors: $\frac{27a^2 - 3w}{6a^3} = \frac{3(9a^2 - w)}{6a^3} = \frac{9a^2 - w}{2a^3}$.

795. **b.** Substitute 1 for x in the function: $x^2 - 4x + 1 = (1)^2 - 4(1) + 1 = 1 - 4 + 1 = -3 + 1 = -2$.

796. **a.** Since the distance from the wall is known, the formula would be $\frac{x}{5} + 2 = 10$. To find x, start by subtracting the 2, giving 8; then, $\frac{x}{5} = 8$, and $8 \times 5 = 40$; therefore, $x = 40$.

797. **b.** The problem is to find J, the number of Jane's toys. It is given that $J = W + 5$, $W = \frac{1}{3}T$, $T = 4E$, and $E = 6$. Substitute: $T = 4(6) = 24$, $W = \frac{1}{3}(24) = 8$, and $J = 8 + 5$. Thus, $J = 13$ toys.

798. **b.** Let x equal the number of hours to paint the sign if both worked together. In one hour Susan can do $\frac{1}{6}$ of the job, while Janice can do $\frac{1}{5}$ of the job. In terms of x, this becomes: $\frac{1}{6}x + \frac{1}{5}x = 1$ (where 1 represents 100% of the job). Solve for x: $(\frac{1}{6} + \frac{1}{5})x = 1$ or $(\frac{5}{30} + \frac{6}{30})x = 1$. Simplify: $\frac{11}{30}x = 1$ or $x = \frac{30}{11}$. Thus, $x = 2.73$ hours.

799. **b.** First, get the inequality into slope-intercept form by dividing both sides by 5. The inequality becomes $y \geq 2x - 3$. The slope of the line is 2 and the y-intercept is −3. In addition, the line should be a solid line and shaded above the line. This is true for choice **b**.

800. **a.** Substitute the second equation for x in the first equation to get $y + 2(3y + 5) = 3$. Use the distributive property and combine like terms: $y + 6y + 10 = 3$; $7y + 10 = 3$. Subtract 10 from both sides of the equation; $7y + 10 - 10 = 3 - 10$; $7y = -7$. Divide each side of the equal sign by 7 to get $y = -1$. Substitute −1 for y in either original equation to get $x = 2$. The solution is $(2, -1)$.

Set 51 (Page 120)

801. **c.** If two linear equations have the same slope, they are parallel. When the equations in choice c are transformed to slope-intercept form, they become $y = x - 7$ and $y = x + 2$. The coefficient of x in both cases is 1; therefore, they have the same slope and are parallel.

802. **d.** Let I equal the amount of interest earned. Substitute: $I = (300)(7\frac{3}{4}\%)(\frac{30}{12})$. Simplify: $I = (300)(0.0775)(2.5)$. Thus, $I = \$58.13$.

803. **b.** You must write the problem as an equation: $J = 6K$ and $J + 2 = 2(K + 2)$, so $6K + 2 = 2K + 4$, which means $K = \frac{1}{2}$; $J = 6K$, or 3.

804. **c.** $M = 3N$; $3N + N = 24$, which implies that $N = 6$ and $M = 3N$, so $M = 18$. If Nick catches up to Mike's typing speed, then both M and N will equal 18, and then the combined rate will be $18 + 18 = 36$ pages per hour.

805. **b.** Let x equal the percent of students who are failing. The percentages must add up to 100. Therefore, $13\% + 15\% + 20\% + 16\% + x = 100\%$, or $x = 100 - 64$. So $x = 36\%$.

806. **b.** The figure is a parabola in the form $y = x^2$ that crosses the y-axis at $(0,-2)$. Therefore, the equation is $y = x^2 - 2$.

807. **a.** If the mixture is $\frac{2}{3}$ raisins, it must be $\frac{1}{3}$ nuts, or $4(\frac{1}{3}) = 1.3$ pounds.

808. **d.** To solve, rearrange, convert units to feet, and then plug in the values: $A = (f \text{ times } D) \div 1 = (0.5 \times 3000) \div 0.25$; $A = 6,000$ feet.

809. **b.** The denominator $x^2 - 2x - 15$ factors to $(x - 5)(x + 3)$. Multiply each side of the equation by the least common denominator of $(x - 5)(x + 3)$, canceling common factors. This results in the equation $2(x + 3) + 3(x - 5) = 11$. Use the distributive property and combine like terms: $2x + 6 + 3x - 15 = 11$; $5x - 9 = 11$. Add 9 to both sides of the equal sign: $5x = 20$. Divide both sides of the equation by 5 to get $x = 4$.

810. **b.** Let x equal the number of pages in last year's directory; 114 less than twice as many pages as last year's means $2x - 114$, so the equation becomes $2x - 114 = 596$, or $2x = 710$. Thus, $x = 355$ pages.

811. **b.** The problem is to find M, the total number of miles traveled. You are given $D1 = 300$ miles;

$D2 = \frac{2}{3}(D1)$; $D3 = \frac{3}{4}(D1 + D2)$. You know that $M = D1 + D2 + D3$. Substitute: $M = 300 + \frac{2}{3}(300) + \frac{3}{4}[300 + \frac{2}{3}(300)]$. Simplify: $M = 300 + 200 + 375$. Thus, $M = 875$ miles.

812. **b.** You are seeking F, the number of gallons of water in the barrel after the thunderstorm. This final amount of water equals the initial amount plus the added amount, or $F = I + A$. You know $I = 4$ gallons, and using the formula $A = rate \times time$, we solve for $A = (\frac{6 \text{ gallons}}{\text{day}})(\frac{1}{3} \text{ day}) = 2$ gallons. Substitute: $F = 4 + 2 = 6$ gallons.

813. **c.** Consecutive odd integers are positive or negative whole numbers in a row that are two apart, such as 1, 3, 5 or $-23, -21, -19$. To find three consecutive odd integers whose sum is 117, divide 117 by 3 to get 39; $39 - 2 = 37$ and $39 + 2 = 41$. To check, add the three integers: $37 + 39 + 41 = 117$.

814. **c.** You know the three species of songbirds total 120, or $A + B + C = 120$. You know $A = 3B$ and $B = \frac{1}{2}C$. This means $A = \frac{3}{2}C$. Substitute and solve for C: $\frac{3}{2}C + \frac{1}{2}C + C = 120$; $3C = 120$; $C = 40$.

815. **a.** Two equations must be used: $2B + 2H = 32$ and $B = 7H$. This gives $14H + 2H = 32$; $16H = 32$; $H = 2$, and $B = 14$; $A = B \times H$; $A = 2 \times 14 = 28$ square meters.

816. **c.** The figure is a parabola opening down, so the equation is in the form $y = -x^2$. Since the graph crosses the y-axis at $+1$, the equation is $y = -x^2 + 1$.

► Section 6—Geometry

Set 52 (Page 124)

817. **c.** A trapezoid by definition is a quadrilateral with exactly one pair of parallel sides.

818. **b.** A cube has four sides, a top, and a bottom, which means that it has six faces.

819. **a.** A polygon is a plane figure composed of three or more lines.

820. **d.** An acute angle is less than 90°.

821. **a.** A straight angle is exactly 180°.

822. **c.** A right angle is exactly 90°.

823. **b.** Because parallel lines never intersect, choice **a** is incorrect. Perpendicular lines do intersect, so choice **c** is incorrect. Choice **d** is incorrect because intersecting lines have only one point in common.

824. **a.** A triangle with two congruent sides could either be isosceles or equilateral. However, because one angle is 40°, it cannot be equilateral (the angles would be 60°).

825. **c.** The sum of the angles on a triangle is 180°. The two angles given add to 90°, showing that there must be a 90° angle. It is a right triangle.

826. **a.** All of the angles are acute, and all are different. Therefore, the triangle is acute scalene.

827. **b.** Both the isosceles trapezoid and the square have congruent diagonals, but only the square has diagonals that are both congruent and perpendicular.

828. **d.** The three angles of a triangle add up to 180°. When you subtract 42 and 59 from 180, the result is 79°.

829. **d.** Squares, rectangles, and rhombuses are quadrilateral (have four sides), and each has two pairs of parallel sides. However, all angles in both squares and rectangles are 90°. Therefore, only a rhombus could contain two angles that measure 65°.

830. **c.** If the pentagons are similar, then the sides are in proportion. Because AB is similar to FG, and $AB = 10$ and $FG = 30$, the second pentagon is three times as large as the first pentagon. Therefore, HI is three times as large as CD, which gives a length of 15.

831. **c.** The greatest area from a quadrilateral will always be a square. Therefore, a side will be 24 ÷ 4 = 6 feet. The area is $6^2 = 36$ square feet.

832. **b.** The area of the square is $4 \times 4 = 16$ square feet; the area of the circle is $\pi(2^2) = \pi4$. The difference is $16 - 4\pi$.

► Set 53 (Page 126)

833. **c.** The perimeter is $4 \times 4 = 16$ (or $4 + 4 + 4 + 4$) for the square, and π times the diameter of 4 for the circle. This is a difference of $16 - 4\pi$.

834. **a.** In order to find the amount of fencing, the perimeter needs to be determined: $120 + 120 + 250 + 250 = 740$ feet.

835. **a.** Use the Pythagorean theorem: $a^2 + b^2 = c^2$. The hypotenuse is found to be 5: $3^2 + 4^2 = 9 + 16 = 25$, and the square root of 25 is 5. The sum of the sides is the perimeter: $3 + 4 + 5 = 12$.

836. **d.** There are four sides measuring 4, and two sides measuring 8. Therefore, the perimeter is $(4 \times 4) + (2 \times 8) = 32$.

837. **c.** The perimeter is the sum of the triangle's two legs plus the hypotenuse. Knowing two of the sides, you can find the third side, or hypotenuse (h), using the Pythagorean theorem: $a^2 + a^2 = h^2$, which simplifies to $2a^2 = h^2$. So $h = \sqrt{2a^2}$. This means the perimeter is $2a + \sqrt{2a^2}$.

838. **c.** To find the perimeter, you can double each of the sides and add the sums: $2(834) + 2(1,288) = 1,668 + 2,576 = 4,244$ feet.

839. **c.** To find the area, multiply the length by the width: $10 \times 6.5 = 65$. The area is 65 square feet.

840. **d.** The rectangular portion of the doorway has two long sides and a bottom: $(2 \times 10) + 4 = 24$. The arc is $\frac{1}{2}\pi d = 2\pi$.

841. **c.** The sum of the side lengths is $7 + 9 + 10 = 26$.

842. **d.** Find the slant height using the Pythagorean theorem: $6^2 + 8^2 = 36 + 64 = 100$. The square root of 100 is 10, so that is the measure of the missing side (the slant height). The perimeter is, therefore, $(2 \times 18) + (2 \times 10) = 56$.

843. **d.** The curved portion of the shape is $\frac{1}{4}\pi d$, which is 4π. The linear portions are both the radius, so the solution is simply $4\pi + 16$.

844. **c.** The perimeter is equal to $4 + 7 + 13 = 24$.

845. **b.** The sum of the sides is the perimeter of the figure. There are four sides that measure 6 inches, so 4×6 is 24. Solve $40 - 24$ to get 16 inches for the remaining two sides. Divide 16 by 2 since the two sides are the same measure. Therefore, $x = 8$ inches.

846. **c.** The perimeter is equal to $(2 \times 4) + (2 \times 9) = 26$.

847. **d.** The angles labeled 120° are alternate interior angles of the lines c and d. When the alternate interior angles are congruent (the same measure), the lines are parallel. Therefore, lines c and d are parallel.

848. **c.** The 60° angle and the angle between angle F and the 90° angle are vertical angles, so this angle must be 60°. The 90° angle is supplementary to angle F and the adjacent 60° angle. $180° - 90° - 60° = 30°$.

Set 54 (Page 129)

849. **b.** Vertical angles are non-adjacent angles formed by two intersecting lines, such as angles 1 and 3 and angles 2 and 4 in the diagram. When two lines intersect, the vertical angles formed are congruent. Therefore, angles 1 and 3 are congruent.

850. **d.** In order to form a triangle, the sum of the two shortest sides must be greater than the longest side. In choice **d**, the two shortest sides are 2 and 2; $2 + 2 = 4$, which is greater than the largest side of 3.

851. **b.** The shortest side is opposite the smallest angle. The smallest angle is 44°, angle ABC. Therefore, the shortest side is AC.

852. **d.** If two angles are 60°, the third must also be 60°. This is an equilateral triangle. All sides are, therefore, equal.

853. **b.** Because the lines are parallel, the corresponding angles 1 and 5 are congruent. Since angles 5 and 8 are vertical angles, they are also congruent. Therefore, angle 1 is also congruent to angle 8.

854. **c.** The sum of the angles is 180°. The two angles given add to 103°, so angle $ABC = 180 - 103 = 77°$.

855. **a.** The distance between Plattville and Quincy is the hypotenuse of a right triangle with sides of length 80 and 60. The length of the hypotenuse equals $\sqrt{80^2 + 60^2}$, which equals $\sqrt{6,400 + 3,600}$, which equals $\sqrt{10,000}$, which equals 100 miles.

856. **a.** The side opposite the largest angle is the longest side. In this case, it is side AB.

857. **a.** The correct answer is 48°.

858. **d.** Supplementary angles add up to 180°. When given one and asked to find its supplement, subtract the given angle from 180°. The correct answer is 96°.

859. **b.** The line right, or 90°, angle is split into two complementary angles. The given angle is 32°; therefore, $90 - 32 = 58°$.

860. **b.** Find the correct answer by dividing the area by the given side length, because area = length × width: 72 ÷ 12 = 6 inches.

861. **a.** The formula for finding the area of triangle is $\frac{1}{2}$ × the base × the height; therefore, if you divide the given area, 60, by the height, 15, it will give half the base, 4. The base is 8 centimeters.

862. **d.** This is the only choice that includes a 90° angle.

863. **c.** The 135° angle and its adjacent angle within the triangle are supplementary, so 180 − 135 = 45°. Angle *B* and the remaining unknown angle inside the triangle are vertical, so the angle within the triangle's measure is needed: 180 − 60 − 45 = 75°, so angle *B* is also 75°.

864. **d.** If the figure is a regular decagon, it can be divided into 10 equal sections by lines passing through the center. Two such lines form the indicated angle, which includes three of the 10 sections; $\frac{3}{10}$ of 360° is equal to 108°.

Set 55 (Page 132)

865. **c.** *PQ* and *RS* are intersecting lines. The fact that angle *POS* is a 90° angle means that *PQ* and *RS* are perpendicular, indicating that all the angles formed by their intersection, including *ROQ*, measure 90°.

866. **b.** The dimensions of triangle *MNO* are double those of triangle *RST*. Line segment *RT* is 5 centimeters; therefore, line segment *MO* is 10 centimeters.

867. **b.** Adjacent angles are angles that share a common side and vertex but do not overlap. Angles 1 and 4 are the only ones NOT adjacent, or next to, each other.

868. **d.** To find the perimeter of a square, multiply the side by 4: 4 × 18 = 72 square feet.

869. **b.** This shape can be divided into a rectangle and two half circles. The area of the rectangle is 4(8) = 32 square units. The area of each half circle can be found by using the formula $A = \frac{1}{2}\pi r^2$. Since the diameters of the half circles correspond with the sides of the rectangle, use half of these sides for the radii of the circles. For the half circle on top of the figure, the diameter is 8, so the radius is 4; $A = \frac{1}{2}\pi 4^2 = 8\pi$. For the half circle on the side of the figure, the diameter is 4, so the radius is 2; $A = \frac{1}{2}\pi 2^2 = 2\pi$. Therefore, the total area of the figure is 32 + 8π + 2π, which simplifies to 32 + 10π.

870. **c.** An algebraic equation must be used to solve this problem. The shortest side can be denoted *s*. Therefore, $s + (s + 2) + (s + 4) = 24$; $3s + 6 = 24$, and $s = 6$.

871. **a.** The correct answer is 24.84 feet. It is found by dividing the circumference by 3.14.

872. **c.** A regular pentagon is a polygon with five equal sides; therefore, 90 ÷ 5 = 18 feet, the length of each side.

873. **d.** First, the radius needs to be found, which is $\frac{1}{2}$ of the diameter: *r* = 47 centimeters; then, to find the area, square the radius and multiply by π. The correct answer is 2,209π.

874. **c.** In order to find the perimeter, the hypotenuse of the triangle must be found. This comes from recognizing that the triangle is a 5-12-13 triangle or by using the Pythagorean theorem. Therefore, 5 + 12 + 13 = 30.

875. **c.** *DE* is 2.5 times greater than *AB*; therefore, *EF* is 7.5 and *DF* is 10. Add the three numbers together to arrive at the perimeter.

876. **b.** Since the triangle is a right isosceles, the nonright angles are 45°.

877. **a.** To get the height of the triangle, use the Pythagoreon theorem: $6^2 + \text{height}^2 = 10^2$. The height equals 8. Then 5 is plugged in for the base and 8 for the height in the area equation $A = \frac{bh}{2}$, which yields 20 square units.

878. **c.** The sum of the measures of the interior angles of a hexagon is 720. In a regular hexagon, each of the six interior angles has the same measure. Therefore, the measure of one angle is $720 \div 6 = 120°$.

879. **b.** Since the five-inch side and the 2.5-inch side are similar, the second triangle must be larger than the first. The two angles without congruent marks add up to 100°, so $180 - 100 = 80°$. This is the largest angle, so the side opposite it must be largest, in this case, side *b*.

880. **c.** If angle 1 is 30°, angle 3 must be 60° by right triangle geometry. Because the two lines are parallel, angles 3 and 4 must be congruent. Therefore, to find angle 5, angle 4 must be subtracted from 180°. The answer is 120°.

Set 56 (Page 135)

881. **b.** The sum of the interior angles of any triangle must add to 180°. Also, the angles of the base of the triangle (interior angles *B* and *C*) are supplementary to the adjacent exterior angles. This means interior angle $B = 180 - 80 = 100°$, and interior angle $C = 180 - 150 = 30°$. Thus, angle *A* must equal $180 - 130 = 50°$.

882. **c.** In a parallelogram, consecutive angles are supplementary. Therefore, the measure of angle *ABC* is equal to $180 - 88 = 92°$.

883. **d.** Because one base angle of the isosceles triangle is 70°, the other must be 70° also. Therefore, since the sum of the interior angles of any triangle is 180°, the vertex angle is $180 - 140 = 40°$.

884. **d.** The diameter of the fan blade will be 12 inches. The circumference is πd, giving a circumference of 12π inches.

885. **c.** The equation is $2\pi r = \frac{1}{2}\pi r^2$. One *r* term and both π terms cancel to leave $2 = \frac{1}{2}r$. Multiply both sides of the equal sign by 2 to get $r = 4$.

886. **b.** The correct answer is 13.56π square centimeters.

887. **a.** The top rectangle is 3 units wide and 3 long, for an area of 9. The bottom rectangle is 10 long and 23 wide, for an area of 230. These add up to 239.

888. **c.** The perimeter is the total of the length of all sides. Since the figure is a rectangle, opposite sides are equal. This means the perimeter is $5 + 5 + 2 + 2 = 14$.

889. **c.** $5 \times 3 \times 8$ is 120; $120 \div 3 = 40$.

890. **b.** Similar figures grow and shrink in proportion to each other. The larger square was achieved by squaring the area of the smaller square. The length of the sides of the larger square is 81 inches.

891. **a.** The correct answer is 21 feet.

892. **c.** Each nine-foot wall has an area of 9×8, or 72 square feet. There are two such walls, so those two walls combined have an area of 72×2, or 144 square feet. Each 11-foot wall has an area of 11×8, or 88 square feet, and again there are two such walls: 88 times 2 equals 176. Finally, add 144 and 176 to get 320 square feet.

893. **c.** Multiply the given dimensions to get a product, or area, of 1,125 square inches.

894. **d.** Take the square root of 324 to find the length of one side of the square; therefore, the correct answer is 72 feet.

895. **b.** The surface area of the walls is four walls of 120 square feet. This gives 480 square feet. The area of the door and window to be subtracted is 12 + 21 square feet = 33 square feet. Therefore, 447 square feet are needed. This would be 5 rolls.

896. **c.** The area is $\frac{1}{2}$ *base* × *height*. This gives $\frac{1}{2}(4)$ (8) = 16.

Set 57 (Page 138)

897. **a.** The area of a parallelogram can be found by multiplying the base times the height, or $A = bh$. The base of the figure is 12, even though it appears at the top of the diagram and the height of the figure is 4. The area is 12 × 4 = 48 square units.

898. **b.** Area is equal to base times height: 2 × 4 = 8 square units.

899. **a.** Divide the perimeter by 4 in order to find the side length, which is 17; then square that number to find the area of the square mirror: 289 square inches.

900. **b.** This must be solved with an algebraic equation: L = 2W + 4; 3L + 2W = 28. Therefore, 6W + 12 + 2W = 28; 8W = 16; W = 2; L = 8. 2 × 8 = 16 square inches.

901. **c.** The longest object would fit on a diagonal from an upper corner to an opposite lower corner. It would be in the same plane as one diagonal of the square base. This diagonal forms one side of a right triangle that has a hypotenuse the length of the longest object that can fit in the box. First, use the Pythagorean theorem to find the length of the diagonal of the square base: $3^2 + 3^2 = c^2$; $c = \sqrt{18}$. Next, find the length of the second side, which is the height of the box. Since the square base has an area of 9, each side must be 3 feet long; we know the volume of the box is 36

cubic feet, so the height of the box (h) must be $3 \times 3 \times h = 36$, or $h = 4$ feet. Finally, use the Pythagorean theorem again to find the hypotenuse, or length of the longest object that can fit in the box: $\sqrt{18}^2 + 4^2 = h^2$; 18 + 16 $= h^2$; $h = \sqrt{34} = 5.8$ feet.

902. **b.** The lot will measure 200 feet by 500 feet, or 100,000 square feet in all (200 feet × 500 feet). An acre contains 43,560 square feet, so the lot contains approximately 2.3 acres (100,000 ÷ 43,560). At $9,000 per acre, the total cost is $20,700 (2.3 × $9,000).

903. **b.** If the circle is 100π square inches, its radius must be 10 inches, using the formula $A = \pi r^2$. Side AB is twice the radius, so it is 20 inches.

904. **d.** Each quilt square is $\frac{1}{4}$ of a square foot; 6 inches is $\frac{1}{2}$ of a foot, 0.5 × 0.5 = 0.25 of a square foot. Therefore, each square foot of the quilt requires 4 quilt squares; 30 square feet × 4 = 120 quilt squares.

905. **c.** There is a right triangle of hypotenuse 10 and a leg of 6. Using the Pythagorean theorem, this makes the height of the rectangle 8. The diagram shows that the height and width of the rectangle are equal, so it is a square; 8^2 is 64.

906. **a.** The Pythagorean theorem states that the square of the length of the hypotenuse of a right triangle is equal to the sum of the squares of the other two sides, so you know that the following equation applies: $1^2 + x^2 = 10$, so $x^2 = 10 - 1 = 9$, so $x = 3$.

907. **b.** The triangles inside the rectangle must be right triangles. To get the width of the rectangle, use the Pythagorean theorem: $3^2 + x^2 = 18$. $x = 3$, and the width is twice that, or 6. This means the area of the shaded figure is the area of the rectangle minus the area of all four triangles. This is $12 \times 6 - 4 \times (\frac{1}{2} \times 3 \times 3)$. This equals 54.

908. **b.** Since the canvas is 3 inches longer than the frame on each side, the canvas is 31 inches long (25 + 6 = 31) and 24 inches wide (18 + 6 = 24). Therefore, the area is 31 × 24 inches, or 744 square inches.

909. **d.** Find the approximate area by finding the area of a circle with a radius of 4 feet, which is the length of the leash: $A = \pi r^2$, so $A = 3.14(4)^2$ = 3.14(16) = 50.24 ≈ 50 square feet.

910. **b.** The formula for finding the area of a triangle is $Area = \frac{1}{2}(base \times height)$. In this case, area $= \frac{1}{2}(7.5 \times 5.5)$ or 20.6 square inches.

911. **a.** Since the area of one tile is 12 inches × 12 inches = 144 square inches, which is one square foot, one tile is needed for each square foot of the floor. The square footage of the room is 10 × 15 = 150 square feet, so 150 tiles are needed.

912. **b.** The correct answer is 12 inches.

Set 58 (Page 141)

913. **b.** The height times the length gives the area of each side. Each side wall is 560 square feet; multiply this times 2 (there are two sides). The front and back each have an area of 320 square feet; multiply this times 2. Add the two results to get the total square feet: 20 × 28 = 560 × 2 = 1,120; 20 × 16 = 320 × 2 = 640; 640 + 1,120 = 1,760. The total square footage is 1,760. Each gallon covers 440 square feet: 1,760 ÷ 440 = 4.

914. **d.** If the shorter sides are each 4 centimeters, then the longer sides must each equal (40 − 8) ÷ 2; therefore, the length of each of the longer sides is 16 centimeters.

915. **a.** Take the square root of 676 to find the length of each side; therefore, each side measures 26 inches.

916. **d.** The total area of the kitchen (11.5 × 11.5) is 132.25 square feet; the area of the circular inset ($\pi \times 3^2 = 9\pi$) is 28.26 square feet: 132.25 − 28.26 = 103.99, or approximately 104 square feet.

917. **c.** To find the size of the piece of glass needed, it is necessary to find the area of the space: 18 × 32 = 576 square feet.

918. **a.** The area of each poster is 864 square inches (24 inches × 36 inches). Kari may use four posters, for a total of 3,456 square inches (864 × 4). Each picture has an area of 24 square inches (4 × 6); the total area of the posters should be divided by the area of each picture, or 3,456 ÷ 24 = 144.

919. **d.** The area of the picture is 475 square inches (25 × 19), and the area of the frame is 660 square inches. The difference—the area the mat needs to cover—is 185 square inches.

920. **a.** Divide the area, 54, by $\frac{1}{2}$ of the base in order to find the height of the triangle: 54 ÷ 9 = 6 centimeters.

921. **d.** The formula for finding the length of fabric is $Area = width \times length$; 54 inches is equal to $1\frac{1}{2}$ yards, so in this case the equation is 45 = 1.5 × $length$, or 45 ÷ 1.5 = $length$ or 30 yards.

922. **a.** The total area of the long walls is 240 square feet (8 × 15 × 2), and the total area of the short walls is 192 square feet (8 × 12 × 2). The long walls will require two cans of Moss and the short walls will require one can of Daffodil.

923. **d.** Find the area of the room, then multiply it by the price per square foot: 23.50(14 × 20) = 23.50(280) = 6,580. The correct answer is $6,580.

924. **c.** First, convert a tile's inches into feet. Each tile is 0.5 foot. The hallway is 10 feet wide, but since the lockers decrease this by two feet, the width of the tiled hallway is eight feet. To get the area, multiply 72 by 8, which is 576 feet. Then divide that by the tile's area, which is 0.5 foot × 0.5 foot, or 0.25 foot; $576 \div 0.25 = 2{,}304$ tiles.

925. **c.** The correct answer is 18 inches.

926. **c.** Since the waiter folded the napkin in half to get a five-by-five-inch square, the unfolded napkin must be 10 inches by five inches. Area is length times width, which in this case is 50 square inches.

927. **d.** The area of the dark yard is the area of her yard minus the circle of light around the lamp. That is $20^2 - \pi \times 10^2$, or $400 - 100\pi$.

928. **d.** Because the radius of the half circle is 3, and it is the same as half the base of the triangle, the base must be 6. Therefore, the area of the triangle is $\frac{1}{2}bh = 12$. The area of the circle is πr^2, which is equal to 9π. Therefore, the half circle's area is $\frac{9\pi}{2}$. Adding gives $\frac{9\pi}{2} + 12$.

Set 59 (Page 143)

929. **b.** The total area of the stores is $20 \times 20 \times 35 = 14{,}000$ square feet. The hallway's area is 2,000 feet. The total square footage is 16,000.

930. **a.** The surface area of the trunk can be found by finding the sum of the areas of each of the six faces of the trunk. Since the answer is in square feet, change 18 inches to 1.5 feet: $2(4 \times 2) + 2(4 \times 1.5) + 2(2 \times 1.5) = 2(8) + 2(6) + 2(3) = 16 + 12 + 6 = 34$. Subtract the area of the brass ornament: $34 - 1 = 33$ square feet.

931. **c.** To find the area of the given circle, square 7 and multiply by π. The correct answer is 49π.

932. **a.** Divide the circumference by 3.14 to find the diameter; therefore, the correct answer is 42 inches.

933. **d.** To get the surface area of the walls, the equation is circumference times height. To get the circumference, plug 10 feet into the equation $C = 2\pi r$. This gets a circumference of 20π. When this and the height of 8 feet are plugged into the surface area equation, the answer is 160π square feet.

934. **d.** The perimeter is equivalent to the distance around. Find the perimeter by adding all of the sides: $75 + 75 + 150 + 150 = 450$ meters.

935. **a.** The area of a triangle is $A = \frac{1}{2}(b \times h)$. Since $b = 2h$, you have $16 = \frac{1}{2}(2h)(h)$ or $h^2 = 16$; $h = 4$ inches.

936. **b.** The correct answer is 15 centimeters.

937. **a.** $Area = \frac{1}{2}(b \times h)$. To get the height of the triangle, use the Pythagorean theorem: $3^2 + height^2 = 5^2$, so height = 4. When this is plugged into the area equation, you'll get an area of 6 square units for half of the triangle. Double this, and the answer is 12 square units.

938. **c.** The Pythagorean theorem gives the height of the parallelogram, 4; the area is $8 \times 4 = 32$.

939. **b.** Since the top and bottom are parallel, this figure is a trapezoid whose $area = (top + bottom)(height) \div 2$. After plugging in the values given, you have $(3 + 6)(10) \div 2 = 45$ square units.

940. **c.** The Pythagorean theorem is used to find the length of the diagonal: $5^2 + 12^2 = 169$. The square root of 169 is 13.

941. **d.** The shaded area is the difference between the area of the square and the circle. Because the radius is 1, a side of the square is 2. The area of the square is 2×2, and the area of the circle is $\pi(1^2)$. Therefore, the answer is $4 - \pi$.

942. **b.** The area of the rectangle is 5(4); the area of the triangle is $\frac{1}{2}(3)(4)$. The sum is 20 + 6 = 26.

943. **a.** There is no difference between the square and parallelogram in area. The formula for both is *base times height*. Both have areas of 16 square inches.

944. **d.** The correct answer is 36π.

Set 60 (Page 146)

945. **c.** A circle is 360°, so 40° is one-ninth of a circle. Multiply the perimeter of the track, 360 feet, by one-ninth, to get the answer of 40 feet.

946. **a.** The formula for finding circumference is πd, or C = 3.14 × 2.5. The circumference is 7.85 inches.

947. **d.** The longest path will be a circular path around the post. This will be equal to the circumference of the circle, which is 2πr. This gives 40π feet.

948. **a.** The formula for determining the area is $A = \pi r^2$, where r is the radius of the tabletop. Since the diameter is 36 inches, the radius is half of 36, or 18 inches. Using the formula, A = 3.14(18)² = 3.14(324) = 1,017.36.

949. **d.** The formula for circumference is 2πr. In this case, the equation is 2 × 3.14 × 4.5 = 28.26 or $28\frac{1}{4}$ inches.

950. **b.** The formula for finding the diameter—the minimum length a spike would need to be—is C = πd, in this case, 43.96 = 3.14 × d; 43.96 ÷ 3.14 = 14.

951. **b.** The formula for determining the circumference of a circle is C = 2πr, in this case, C = 2 × 3.14 × 25 or 157.

952. **a.** The formula for finding the diameter is *Circumference = the diameter × π* (C = πd): 18.84 = πd, or 18.84 = π(6). The diameter is 6.

953. **a.** To find the radius, take the square root of the coefficient, 49; therefore, 7 is the correct answer.

954. **a.** The area of the dough is 216 square inches (18 × 12). To find the area of the cookie cutter, first find the radius. The formula is C = 2πr, in this instance, 9.42 = 2 × 3.14 × r, or 9.42 = 6.28 × r. Divide 9.42 by 6.28 to find r, which is 1.5. The formula for area of a circle is (πr²), in this case, 3.14 × 1.5², or 7.07 square inches. So, the area of the cookie cutter circle is 7.07 square inches. Divide 216 (the area of the dough) by 7.07 (the area of the cookie cutter), and the result is 30.55 or approximately 31 cookies.

955. **b.** If the wheel has a diameter of 27 inches, the circumference is π × 27. Multiplied by 100 turns, the distance Skyler biked is 2,700π inches.

956. **a.** An isosceles triangle has two sides that measure the same and one that measures different. Add them together to find the perimeter: 8 + 14 + 14 = 36 centimeters.

957. **c.** A regular octagon has eight equal sides; therefore, the perimeter equals 8 × 13. The correct answer is 104 centimeters.

958. **b.** The perimeter of the room is 36 feet (9 × 4); 36 ÷15 (the length of each garland) = 2.4. So Bridget will need three garlands.

959. **d.** Through trial and error, find dimensions that, when their sum is doubled, give you 38. The correct answer is 10 centimeters × 9 centimeters.

960. **a.** The unpainted section of the cloth is $6\frac{2}{3}$ feet by $8\frac{2}{3}$ feet, because each side is shortened by 16 inches (an eight-inch border at each end of the side). To find the perimeter, add the length of all four sides, or 6.66 + 6.66 + 8.66 + 8.66 = 30.66 or $30\frac{2}{3}$.

Set 61 (Page 148)

961. **b.** Find the area of the pillow: $2.5 \times 1.5 = 3.75$ square feet.

962. **d.** The height of the triangle using the Pythagorean theorem, $10^2 + height^2 = 20^2$, is equal to $10\sqrt{3}$. Using the area formula for a triangle, $A = \frac{1}{2}(20)(10\sqrt{3}) = 100\sqrt{3}$. To find the amount of water the fish tank can hold, or the volume, multiply the area of the base by the depth of the tank: $(100\sqrt{3})(10) = 1,000\sqrt{3}$.

963. **b.** Facing in the exact opposite direction is turning through an angle of 180°. Therefore, the new compass reading will be 32° south of west. This is equivalent to the complementary angle $(90 - 32) = 58°$ west of south.

964. **d.** Since the ceiling and the floor are parallel, the acute angles where the stairs meet the ceiling and floor are equal. These are also supplementary to the obtuse angles, so $180 - 20 = 160°$.

965. **d.** A 90° angle is called a right angle.

966. **b.** When the sum of two angles is 180°, the angles are supplementary or supplemental to each other. To find the supplement, subtract 35° from 180°, or $180 - 35 = 145°$.

967. **a.** Since the bumpers are parallel, then the angle whose measure is needed must be equal to the angle in between the two equal angles, or 36°.

968. **a.** The ladder makes a 54° angle with the ground, and the angle between the ground and the house is 90°. Since there are 180° in any triangle, $180 - 54 - 90 = 36°$ between the ladder and the house.

969. **d.** For the ramp to go up 8 inches, it must have a horizontal distance of 8 feet or 96 inches. Using the Pythagorean theorem, the diagonal distance $D^2 = 8^2 + 96^2 = 64 + 9,216 =$ 9,280. Since $9,280 = 4 \times 2,320$, $\sqrt{9,280} = \sqrt{4} \times \sqrt{2,320} = 2\sqrt{2,320}$. We can do this twice more to get $2\sqrt{2,320} = 2\sqrt{4} \times \sqrt{580} = 2 \times 2 \times \sqrt{4} \times \sqrt{145} = 8\sqrt{145}$.

970. **c.** The vertex is the point where the two rays of an angle meet.

971. **c.** The triangle created by the chicken, the cow, and the bowl of corn is a right triangle; the 90° angle is at the point where the chicken is standing. Use the Pythagorean theorem ($a^2 + b^2 = c^2$) to find the missing side of this right triangle. In this case, $a = 60 \times 60$; $c = 80 \times 80$ (the hypotenuse is c), or $3,600 + b^2 = 6,400$. To solve, $6,400 - 3,600 = 2,800$; the square root of 2,800 is 52.915, rounded to 53. The chicken will walk about 53 feet to cross the road.

972. **c.** Since the ships are going west and north, their paths make a 90° angle. This makes a right triangle where the legs are the distances the ships travel, and the distance between them is the hypotenuse. Using the Pythagorean theorem, $400^2 + 300^2 = distance^2$. The distance is 500 miles.

973. **b.** The measure of the arc of a circle is the same measure as the central angle that intercepts it.

974. **d.** The total number of degrees around the center is 360. This is divided into six equal angles, so each angle is determined by the equation: $360 \div 6 = 60$.

975. **c.** An angle that is more than 90° is an obtuse angle.

976. **c.** The Pythagorean theorem is used to solve this problem. Forty feet is the hypotenuse, and 20 feet is the height of the triangle: $40^2 = 20^2 + w^2$; $w^2 = 1,200$; $w = 34.6$. This is closest to 35 feet.

Set 62 (Page 151)

977. d. The height of the tree and the length of its shadow are in proportion with Lara's height and the length of her shadow. Set up a proportion comparing them. Let x equal the height of the tree: $\frac{x}{40} = \frac{5.5}{10}$. Cross multiply to get $10x = 220$. Therefore, $x = 22$ feet.

978. a. The angle that the ladder makes with the house is 75°, and the angle where the house meets the ground must be 90°, since the ground is level. Since there are 180° in a triangle, the answer is $180 - 90 - 75 = 15°$.

979. b. To bisect something, the bisecting line must cross halfway through the other line; $36 \div 2 = 18$ feet.

980. d. First, find the number of yards it takes Mario to go one way: $22 \times 90 = 1,980$. Then double that to get the final answer: 3,960.

981. c. First, find the area of the brick wall: $A = lw$, or 10 feet $\times$ 16 feet $= 160$ square feet. Now convert 160 square feet to square inches—but be careful! There are 12 linear inches in a linear foot: but there are 144 (12^2) square inches in a square foot: $160 \times 144 = 23,040$ square inches. Divide this area by the area of one brick (3 inches $\times$ 5 inches $= 15$ square inches): $\frac{23,040}{15} = 1,536$.

982. b. The midpoint is the center of a line. In this case, it is $400 \div 2$, or 200 feet.

983. d. When two segments of a line are congruent, they are the same length. Therefore, the last block is 90 feet long (the same as the second block). To arrive at the total distance, add all the segments together: $97 + 2(90) + 3(110) + 90 = 607$.

984. a. A transversal line crosses two parallel lines. Therefore, if line E traverses line A, it also traverses (crosses) line B. If line E is not

perpendicular to line A, then it will also cross both line C and line D, but this might not happen.

985. c. A perpendicular line crosses another line to form four right angles. This will result in the most even pieces.

986. b. Since the houses all have even house numbers, they are all on the same side of the straight street. This makes them *collinear*, all in a straight line.

987. c. The sum of the measures of the exterior angles of any convex polygon is 360°.

988. b. To find the amount of fencing, you need to find the perimeter of the yard by adding the lengths of all four sides: $60 + 125 + 60 + 125 = 370$. Then, subtract the length of the four-foot gate: $370 - 4 = 366$ feet.

989. b. The third side must measure between the difference and the sum of the two known sides. Since $7 - 5 = 2$ and $7 + 5 = 12$, the third side must measure between 2 and 12 units.

990. b. The volume of the water is $10 \times 10 \times 15 = 1,500$ cubic inches. Subtracting 60 gets the answer, 1,440 cubic inches.

991. c. The formula for volume is $V = l \times w \times h$; therefore, $8.5 \times 5.5 \times 10 = 467.5$ cubic inches.

992. d. To find the volume of a right circular cylinder, use the formula $V = \pi r^2 h$. $V = (3.14)(6)^2(20) = (3.14)(36)(20) = 2,260.8 \approx 2,261$ cubic feet.

Set 63 (Page 154)

993. d. The volume of concrete is 27 cubic feet. The *volume is length times width times depth* (or *height*), or $(L)(W)(D)$, so $(L)(W)(D)$ equals 27. The length L is 6 times the width W, so L equals $6W$. The *depth* is 6 inches, or 0.5 feet. Substitute what you know about the *length* and *depth* into the original equation and solve for W: $(L)(W)(D) = (6W)(W)(0.5) = 27$; $3W^2$ equals 27; W^2 equals 9, so W equals 3. To get the length, remember that L equals $6W$, so L equals $(6)(3)$, or 18 feet.

994. a. The amount of water held in each container must be found. The rectangular box starts with 16 square inches × 9 inches = 144 cubic inches of water. The cylindrical container can hold 44π cubic inches of water, which is approximately 138 cubic inches. Therefore, the container will overflow.

995. b. The volume of the box is $144 + 32 = 176$ cubic inches. That divided by the base of the box gets the height, 11 inches.

996. b. The volume of the briefcase is $24 \times 18 \times 6$ inches, or 2,596 cubic inches. The volume of each notebook is $9 \times 8 \times 1$ inches, or 72 inches. Dividing the volume of the briefcase by the volume of a notebook gets an answer of 36 notebooks.

997. c. Think of the wire as a cylinder whose volume is $(\pi)(r^2)(h)$. To find the length of wire, solve for h, in inches. One cubic foot equals $(12)^3$ cubic inches, or 1,728 cubic inches. Therefore, $(\pi)(0.5^2)(h) = 1,728$; $h = \frac{4(1,728)}{\pi}$; $h = 6,912 \div \pi$.

998. c. The proportions of Daoming's height to his shadow and the pole to its shadow are equal. Daoming's height is twice his shadow, so the pole's height is also twice its shadow, or 20 feet.

999. b. A dilation changes the size of an object using a center and a scale factor.

1000. c. Points B, C, and D are the only points in the same line and are thus also in the same plane.

1001. a. This problem can be solved using the method of Similar Triangles. Assume both objects are perpendicular to the earth's surface. The sun's rays strikes them at an equal angle, forming "similar triangles," with the shadow of each as that triangle's base. The ratio of a triangle's height to its base will be the same for both triangles. Therefore, you need know only three of the four measurements to be able to calculate the fourth. Let c equal the unknown height of the pole. Let the height of the sign, a, equal 8 feet, the length of the sign's shadow, b, equal 3 feet, and the length of the pole's shadow, d, equal 15 feet. By similar triangles, $\frac{a}{b} = \frac{c}{d}$. Substitute: $\frac{8}{3} = \frac{c}{15}$. Cross multiply: $(8)(15) = 3c$. Rearrange: $c = \frac{8(15)}{3}$. Thus, $c = 40$ feet.

NOTES

NOTES

NOTES

NOTES

NOTES

NOTES

NOTES

NOTES

NOTES

NOTES

NOTES

NOTES